DIVI DIGITS

By

JOHN RANDALL BELL, Ph.D.

2014

COPYRIGHT - 2014 by J. Randy Bell

ISBN: 978-0-9910692-0-0

PRINTED IN THE UNITED STATES OF AMERICA

PUBLISHED BY CALVARYSTAND, Tunnel Hill, Georgia, *A Nonprofit Organization, Devoted to the Lord's Work and to the Spread of His Truth*

All Scripture quotations are taken from the Authorized Version of the Bible commanded by James, King of Great Britain, France and Ireland, Defender of the Faith.

All rights reserved. No part of this book may be reproduced or transmitted in any form by any means, electronic, or mechanical without permission in writing from the author. Preachers of the gospel may use any part of this book in preparing and presenting the Gospel of Christ to the people of this world.

To my faithful wife, Mary E. Bell, whose love, companionship and prayers have been a constant fount of encouragement in this earthly walk

And to my late brother, Lawrence Lamar Bell, who was a diligent scholar, dedicated Christian and mathematician

PREFACE

The premise of this work is to reveal the symbolism of numbers in the Bible and to expand the thinking of the reader.

The method of research is to study the verses in the Bible accompanying the number to determine its symbolism. The symbolism of a number is the idea associated with that number in the Bible.

The results are that there is an interesting interrelationship between the numbers of the Bible, especially the prime numbers and twin prime numbers. The reader's faith in a Divine Creator will be increased when a thorough study of this publication is engaged.

CONTENTS

CHAPTER

1. The Divine Design Discovered..................1
2. History of the Biblical Symbolism of Numbers...................8
3. The Number Zero - "Undisciplined"...................10
4. The Number One - "Unity"...................18
5. The Number Two - "Witness" 21
6. The Number Three - "Complete"...................24
7. The Number Four - "Humanity"...................29
8. The Number Five - "Grace"...................33
9. The Number Six - "Man"...................35
10. The Number Seven - "Perfect"...................38
11. The Number Eight - "New Beginning"...................44
12. The Number Nine - "Divine Order"...................47
13. The Prime Numbers...................50
14. The Number Ten - "Order"...................52
15. The Number Eleven - "Judgment"...................55
16. The Number Twelve - "Divine Authority"...................58
17. The Number Thirteen - "Rebellion"...................62
18. The Number Fourteen - "Salvation"...................65
19. The Number Fifteen - "Rest"...................68
20. The Number Sixteen - "Love"...................73

21. The Number Seventeen - "Victory"....................................78

22. The Number Eighteen - "Bondage"...................................83

23. The Number Nineteen - "Wisdom"....................................87

24. The Twin Prime Numbers...92

25. The Number Twenty - "Redemption"..................................95

26. The Number Twenty-one - "Evil"......................................99

27. The Number Twenty-two - "Light"...................................103

28. The Number Twenty-three - "Death"................................106

29. The Number Twenty-four - "Priesthood"...........................112

30. The Number Twenty-five - "Forgiveness".........................118

31. The Number Twenty-six - "The Gospel"...........................122

32. The Number Twenty-seven - "Hope"................................126

33. The Number Twenty-eight - "Atonement".........................131

34. The Number Twenty-nine - "Sanctification"....................136

35. The Number Thirty - "Life (Blood) of Christ"...............140

36. The Number Thirty-one - "Cleansing"............................145

37. The Number Thirty-two - "Covenant".............................153

38. The Number Thirty-three - "Purification"......................160

39. The Number Thirty-four - "Spirit-filled".....................165

40. The Number Thirty-five - "Forsaken"............................172

41. The Number Thirty-six - "Deadness".............................178

42. The Number Thirty-seven - "Mighty".............................184

43. The Number Thirty-eight - "Feebleness"........................190

44. The Number Thirty-nine - "Punishment"........................196

45. The Number Forty - "Testing"..203

46. The Number Forty-one - "Leadership"............................211

47. The Number Forty-two - "Condemnation"......................216

48. The Number Forty-three - "Counsel"...............................222

49. The Number Forty-four - "Hell"......................................227

50. The Number Forty-five - "Preservation"..........................231

51. The Number Forty-six - "Legacy"...................................233

52. The Number Forty-seven - "Disappointment"................235

53. The Number Forty-eight - "Sacred"................................237

54. The Number Forty-nine - "Faith"....................................239

55. The Number Fifty - "Holy Spirit"....................................242

56. The Number Sixty - "Strong" 245

57. The Number Seventy - "Fruition"...................................248

58. The Number Eighty - "Strength".....................................252

59. The Number Ninety - "Fruitful".......................................259

60. The Number One Hundred - "Whole"..............................264

61. The Number One Hundred Twenty - "Royalty"..............269

62. The Number One Hundred Fifty - "Earthly Authority"..271

63. The Number One Hundred Fifty-three - "Evangelism"..275

64. The Number Two Hundred - "Unacceptable"..................278

65. The Number Three Hundred - "The Father"......................281

66. The Number Three Hundred Twenty-three - "Dominion" 286

67. The Number Three Hundred Sixty-five - "Consecration" 289

68.. The Number Four Hundred - "Change".........................291

69. The Number Five Hundred - "Exceptional" 294

70. The Number Six Hundred - "Chosen"............................298

71. The Number Six Hundred Sixty-six - "Controller".........300

72. The Divine Design of Digits..303

73. The Designer of All Things...306

CHAPTER 1
THE DIVINE DESIGN DISCOVERED

I. The Crow That Could Count

The rich bottomland yielded an exceptional crop of corn for the last three years. As a result the North Georgia farmer decided to build a larger corn crib. This would not be just any corn crib. He would design and build a steeple on top of it in which to sit, and gaze over his farmland. It was ideal until one day he discovered a crow had plagued the edifice by building a nest in the upper most part. Vexed by the crow, the farmer decided to rid his watchtower of this pest. He repeatedly entered the tower with his shotgun. Each time the crow would position himself to a distant tree and observe the tower. Waiting for a long length of time, the farmer would finally relent and the winged villan would return to claim his nest.

Not to be outwitted by a member of the animal kingdom, the farmer solicited the help of a neighbor. Dressed in the same garb as the farmer with Liberty overalls and a red flannel shirt, the neighbor carrying his shotgun, followed the farmer into the tower, and thirty minutes later, the neighbor exited the building. The farmer remained, and was ready to shoot the intruder. The crow was not deceived for it stayed in the distant tree until the owner within the tower departed. The crow quickly reentered his abode.

The experiment now became a contest. The next day three men with their shotguns entered, wearing identical clothing and two men exited, but the crow was able to count and subtract. The contest continued each day with an increasing number of men, but the crow demonstrated his mathematical ability. Ultimately five men entered the tower, four came out, and went away and the farmer remained inside.

At this point, the crow seemed to have lost count and unable to distinguish between four and five, it returned to its nest in the tower and to its destiny. The culmination of the story has not been recorded, but it is hoped that by the time of the final test, that the farmer gained respect and affection for the crow that could count and subtract. There is a divine Designer who has made the number system and created in animals and in man a knowledge of it.

II. Upon the Shoulders of Giants

The early Greeks influenced by Pythagoras, the father of mathematics, developed a correlation between numbers and concepts. The numeral one was revered as the number of reason. Two, the first even or female number, represented diversity of opinion. Three, the first male number, represented harmony. When asked what a friend was, Pythagoras stated "Another I." The relationships between friends is comparable to the relationship between the divine digits and Biblical ideas.

Archeological evidence in Iraq has revealed that students used clay tablets to study the "Pythagorean Theorem," many years before Pythagoras' name was attached to the theorem. In the days of Abraham in the country of Ur, the study of the right triangle was common. The Pythagorean Theorem is the sum of the square of the two small sides equals the square of the large side of a right triangle, $A^2 + B^2 = C^2$. The achievement of Pythagoras was more of codifying the known mathematical laws instead of discovering them.

In a similar manner, the discovery of the symbolism of numbers in the Bible is more of analyzing the works of previous writers and condensing them to a readable form. This Biblical work has been accomplished upon the shoulders of giants of numerology. In 1974, Dr. Ed. F. Vallowe printed the seventh edition of his book, *Keys to Scripture Numerics*. His publication closely followed the works of E. W. Bullinger, who had written *Number In Scripture* in 1967. Dr. Bullinger credited Dr. Milo Mahan whose work was *Palmoni* which established some of the fundamentals of Biblical numerology. Dr. Milo Mahan in 1863, credited Browne for his writing, *Ordo*

Saeclorum. Dr. Mahan stated that he harmonized with Browne in his discussion of Biblical numbers.

Just as Pythagorus compiled the works of previous unknown mathematicians, this publication is produced from an analysis of previous giants of the faith in Biblical numerology. Care is taken not to copy or plagiarize these previous giants, but to turn the pages of the Bible to find the most correct symbolism of the Biblical numbers. This approach to determine the symbolism of the Biblical numbers is a fresh approach and unique to most researchers of numerical symbolism in the Scriptures. The uncovering of the symbolism of the number zero and of the number nineteen is distinctive in the study of Biblical numbers from other publications. This work has relied upon previous giants in their study of Biblical numbers, but only upon their methods, not upon their findings.

III. The Purity of the Scriptures

The purity and permanency of the Holy Scriptures is clearly stated. It is inconceivable that God would simply place numbers in the Bible in a random manner. From the lips of the wisest man of his day, Solomon, came the words that every word of God is pure (Proverbs 30:5). King David stated also that the Word of God was purified seven times (Psalms 12:6). The Old Testament scribes treated the Torah with great respect and ensured that each copy was exactly like the original, thus maintaining its purity. The Psalmist David declared in Psalm 119:89, "For ever, O Lord, thy word is settled in heaven."

Jesus, the Son of God, said that till Heaven and earth pass, one jot or one tittle shall in no wise pass from the law, till all be fulfilled (Matthew 5:18). He placed emphasis on the fact that every word in the Bible has significant meaning. The Apostle Paul stated that all Scripture is given by inspiration of God and is profitable for instruction (Second Timothy 3:16).

IV. The Historic and Symbolic Value of the Numbers

Every number in the Bible has either a historical or a symbolic value. The Divine Creator has a purpose for every number that appears in the Scriptures. Just as the names of persons and places in the Bible have meaning, the numbers also in the Bible by design of the great Creator have meaning. This study of the Bible is based upon the following: First, some numbers are in scripture to give a historical perspective. The book of Numbers is representative of this fact with all the listings and numberings of the children of Israel by tribes.

Second, some numbers are in scripture solely for their symbolic purpose. The appearance of the number one hundred and fifty-three in the book of St. John puzzles many Bible readers. Why would God place the number of fish that were caught in a net in such a crucial part of the Holy Bible? It clearly has no historical value, yet God placed it in the Scriptures. When the symbolisms of the numbers in the Bible are revealed, then the faith of the reader in the Word of God will increase. There is a divine Creator who designed the digits we use.

Third, some numbers are in scripture to dually provide information about the historical aspect and the symbolic nature of the number. The symbolism of the numbers in the Scriptures will be derived instead of relying totally on lists and the position of an idea or principle in that list, but by studying the meaning of the number in the context of the scriptures in which it is found. Lists can provide almost any meaning to a number in the Bible if one searches extensively. The object of this writing is to strengthen the believer in their most holy faith, and to receive a higher regard for the inspiration of the Holy Bible. Proverbs 25:2 states that it is the "glory of God to conceal a thing: but the honor of kings to search out a matter." God has concealed things in His Word, and it is a royal, honorable task in searching out the symbolism of the numbers in the Holy Scriptures. The study of the digits and their symbolism in the Bible reveals the divine design in the creation of the digits.

The symbolism of a number is the idea associated with that number in the scriptures. Milo Mahan said, "Certain numerals in the Scriptures occur so often in connection with certain classes of ideas, that we are naturally led to associate the one with the other." An examination of the Scriptures where the number is found and searching for a common theme is the method of determining the symbolism of a number.

Milo Mahan also said, "This principle, as I have more than once said, is that of a simple, childlike association of ideas." Difficulty arises when the Bible student searches for the symbolism when the number is mentioned only one or two times in the Holy Scriptures. The childlike association becomes less clear and does not easily appear. In searching for the most appropriate symbolism, the Bible student must fully realize that God is perfect, therefore His works and His words, such as the numbers, must be perfect and perfectly placed in the Holy Scriptures.

The Law of First Mention is one of the many rules which is used in interpreting a portion or word of Scripture. The law is a simple concept for when the first time a word, incident, or phrase occurs in the Holy Scriptures, it gives the key to the meaning when the word is used elsewhere in the Bible. The Law of First Mention is not though consistent in all the Bible.

CHAPTER 2
HISTORY OF THE BIBLICAL SYMBOLISM OF NUMBERS

Associating numbers with virtues or conditions is not new. Saint Jerome (circa. 340-420 A.D.) was one of the greatest early interpreters of the Bible. His translation of the Old Testament from Hebrew into Latin is a classic. It became the Vulgate, or popular version used by the Roman Catholic Church. Along with translating the Bible, Saint Jerome wrote Biblical commentaries which included a symbolic explanation of the Parable of the Sower found in Matthew 13:8. Concerning the hundredfold, the sixtyfold, and the thirtyfold yield from the same soil, Saint Jerome gave a symbolic commentary on this familiar passage.

The method of counting in medieval times was by using the fingers. On the left hand, the number thirty was formed by placing the tip of the index finger on the end of the thumb to form a circle. This was called "the tender embrace." The other fingers of the left hand were outstretched. Saint Jerome stated that thirty is a symbol of marriage, for the joining of the index finger and the thumb represents both husband and wife in an embrace.

For Saint Jerome, sixty symbolized widowhood. The hand position for sixty in medieval times was to place the thumb on the left hand into the palm with the four fingers extended. Sixty was symbolic of widowhood for the widow endured troubles which weighed on her just as the thumb was pressed down into the palm.

The medieval finger sign for the number one hundred was on the right hand. It was the same as the finger position for the number sixty on the left hand. The thumb was depressed into the palm and the other fingers of the right hand were outstretched. Saint Jerome stated that it is more difficult to abstain from the pleasures of this world which were once enjoyed, but the greater will be the reward.

Many cultures associated numbers with different ideas. The idea that numbers stood for some condition or virtue is an old one. Pythagoras, a Greek philosopher and mathematician, taught that numbers were the essence of everything. Pythagoras believed that all things were fittingly ordered according to the nature of numbers. He believed that numbers are the eternal essence. He believed that God is number and number is God.

CHAPTER 3
THE NUMBER ZERO ~ "UNDISCIPLINED"
THE SOURCE

The word "zero" is not found in the Authorized Version of the Bible, but the old English word "nought" is used extensively to portray the meaning of nothing. The first mention of "nought" in the Bible is found in Genesis 29:15 where Laban instructs Jacob that he will not employ him without remuneration. The Bible speaks of the counsel of men and of God being brought to nought as in Nehemiah 4:15, Proverbs 1:25, and Isaiah 8:10. Insinuating that Job served God for the physical rewards he had received from God, Satan answered the Lord with, "Doth Job fear God for nought?" Malachi 1:10 reveals that God had no pleasure in the children of Israel because they would not shut the doors of the temple unless they were paid.

The origin of the word *zero* is from various sources. It probably comes from the Latin form of the word *zephirum* of the Arabic *sifr* which in turn is derived from the Hindu word *sunya.* It means "void" or "empty." The Arabic *sifr* was used in Germany in the thirteenth century by Nemorarious, as *cifra*, from which we have our English word *cipher.*

In the New Testament, Jesus used the word *nought* to describe His final position before the Sanhedrin and Pilate (Mark 9:12). This action was fulfilled in Luke 23:11 when "Herod and his men of war set him at nought and mocked him." In Acts 4:11, Peter equates Christ with the stone which was set at nought of the builders, which is become the head of the corner. Gamaliel advised the Sanhedrin in Acts 5 to refrain from any action against the new sect of followers of Jesus stating, "for if this counsel or this work be of men, it will come to nought."

The Apostle Paul used the word nought to describe the actions of his companions among the Thessalonians. The subject was centered on whether Paul and his company had worked in Thessalonica to receive any rewards. Paul assured them in Second Thessalonians 3:8 that they did not eat any man's bread for nought, but that they had labored and travailed night and day.

The Hebrew word for nought is *chinnam* which means devoid of cost, reason or advantage; without a cause (cost, wages), causeless, to cost nothing, free, innocent, for nothing, in vain. The word nought or naught clearly describes more than the word nothing. It deals with the cost of an item or the wages given to a person. Unfortunately it does mean zero, in that many people labor and receive zero for their efforts.

THE SYMBOLISM

Nought or zero is symbolic of being undisciplined. In Genesis 29:15, Laban realized that after Jacob had resided with him for a month that Jacob should be paid wages so he could control the young man. A person who works for another must have an incentive for which to work. Jacob's incentive was a future bride, Rachel. Jacob would have become undisciplined as a worker unless incentives were established.

This symbolism of being undisciplined is further illustrated by David's example of refusing to freely receive the threshing floor and sacrificial animals of Araunah (Second Samuel 24:24). David demonstrated discipline by paying for the altar and the animals to be used as an offering to God. David showed the same control in his life, when he bought Onan's threshing floor, wooden instruments and animals for an offering to God (First Chronicles 21:24) at the full price.

Satan accused Job of serving God for all the benefits God had given him. Satan misjudged Job in thinking that Job required wages to continue serving God. Satan felt that if Job's wages were nought or zero, then he would not continue serving God. Satan was misinformed of the motive of Job in serving

God (Job 1:9). Satan had attacked an unusual man when his sights focused upon Job.

Solomon wisely stated that the net spread in sight of the bird was done in vain or for nought. The yield would be zero. Clearly the person who hunted in this manner was undisciplined (Proverbs 1:17).

Isaiah warned Israel that they had sold themselves for zero and would be redeemed without money. Because of their undisciplined lifestyle, Israel suffered slavery and domination by their enemies (Isaiah 52:3). An undisciplined or zero type lifestyle produces horrible results. The state of the nation of Israel at the first coming of Christ is a good example of an undisciplined nation.

Clearly the number zero or nought symbolizes undisciplined. A woe is pronounced over the person whose lifestyle of unrighteousness and thievery uses other people. They truly live an undisciplined life without concrete rules whether morally, ethically, or spiritually. (Jeremiah 22:13).

In the days of Malachi, God rebuked Israel stating that no one had the self discipline to shut the doors of the tabernacle or kindle the fires on the altar. They expected to be paid for performing the chores in the tabernacle (Malachi 1:10). The undisciplined or zero person is one who is frivolous in their duties especially towards God. The lack of self control in a person indicates an undisciplined life. The idea associated with nought or zero in the Bible is the concept of being undisciplined either in a person's life or a nation's condition.

THE SERMON

The Undisciplined Person

I. The Frivolity of the Undisciplined - Genesis 29:15

An undisciplined person shows few signs of personal restraint. There is no sense of purpose in their life. Their attitude is one marked with a lack of responsibility. This undesired characteristic sadly can be found in the life of the unsaved or in the life of the saved person.

Thales, a Greek mathematician in the first half of the sixth century B. C., was considered to be one of the "seven wise men" of antiquity. Since the study of mathematics gave little income, Thales, like many mathematicians, was engaged in business ventures. One of his business ventures was extracting salt from a mine on his property. Each day he employed miners to dig the salt, place it on mules, and transport the product to the store at the bottom of the mountain for selling. One day one of the mules stumbled and fell into the small stream which crossed the path of the mules. As the mule rolled over in the stream, the salt dissolved. The mule remembered the experience the next day. The next trip from the salt mine, the mule laid down in the water, and his burden was relieved. The undisciplined would have acted in a frivolous manner by whipping the mule and shouting with a loud voice the next time it laid in the water. Thales was a wise man for the next time that the mule came down the hill, he loaded the mule with bags of sponges. Thales demonstrated self-restraint or discipline which led to the solution of his problem.

The undisciplined person acts in such a frivolous way that his life has no purpose. Laban knew if Jacob was not disciplined and worked for something or someone, then he would have no control over him. The undisciplined person cannot be held to any standards or requirement. The action of Laban proved fruitful, for Jacob was agreeable to work for Laban for the hand of his daughter. You can't make a place for yourself under the sun if you keep sitting in the shade of the family tree.

II. The Fruit of the Undisciplined - Genesis 29:20

The primary fruit of the undisciplined is slothfulness. Most lazy people have about as much initiative as an echo. The slothfulness mentioned in the Bible is an accusation from God who stated that man should work six days and rest one day. No law can be passed which would make a person disciplined in his work habits. The bee that collects the honey doesn't hang around the hive. Self initiative and ambition are good qualities in a person who first must demonstrate discipline.

Sensuality is another fruit of the undisciplined. The person who has no control over his personal life is a zero in the eyes of God. David dropped to his lowest when he chose immoral companionship instead of battling the enemy in the combat zone. Demas showed signs of being very fruitful for the cause of Christ, but abandoned it for the world. The fruit of the undisciplined leads to sorrows.

Sinfulness is the third fruit of the undisciplined. An undisciplined life produces sin. There is no accountability for the person with no order in his life. Many persons have good intentions but never accomplish them because they are undisciplined. A person may have a heart of gold, but so does a hard-boiled egg. To know to do good and do it not is sin. Sin is found not only in an evil act but also in a neglected duty.

III. The Fate of the Undisciplined - Luke 15:13

An undisciplined person is a disappointment to God. Man is the divine creation of God. To live a life without boundaries is living below the potential which God has for the person. There are not enough crutches in the world for all the lame excuses given by one who could live a disciplined life, but does not.

The prodigal son was undisciplined in his lifestyle. He was a great disappointment to his father.

After wasting his life in riotous living, the prodigal son decided upon a future which involved becoming a servant to his father. A son never makes a good transition from sonship to servanthood. He would have become a liability and disability to the father in that position. Eating the shucks with the swine was a disgusting position for a Jewish lad. His feet were ankle deep in mud and swine manure. Truly his life became disgusting to himself.

A disciplined person performs a self-analysis to become a better person whether for God or for society. Ultimately the prodigal son "came to himself" (Luke 15:17) which demonstrated discipline.

CHAPTER 4

THE NUMBER ONE ~ "UNITY"

The Source

The first mention of the number one is found in Genesis 1:9 where the waters were gathered unto one place. In Genesis 2:21, God took one of Adam's ribs and formed Eve. Ephesians 4:4-7 reveals that, "There is one body, and one Spirit, even as ye are called in one hope of your calling. One Lord, one faith, one baptism, One God and Father of all, who is above all, and through all, in you all." The number one is the most common number in the Bible because it is used to designate not only a person in a story but in a calculating, numerical manner.

The Symbolism

The first mention of the number one is found in Genesis 1:9 with the unification of the waters on the earth. The ring Pharaoh (Genesis 41:42) placed on the hand of Joseph symbolized that they were one in government. The father instructed the servant to place a ring on the hand of his prodigal as a sign of unity (Luke 15:22). They were again one in the family. The number one symbolizes unity. In the early years of the New Testament church, the Scriptures mentioned the believers were in one accord. The unity of the church is found in Ephesians 4:4-7 where the unity of the body and the spirit are emblematic of the unity of the hope of the church. There is one Lord, one faith, one baptism, one God and Father of all.

The Sermon

The Importance of One

I. One Man's Disobedience ~ Genesis 3:6

Adam disobeyed God by eating of the fruit offered by his wife, Eve. While Eve was deceived, Adam disobeyed. One man plunged the entire human race into sin by his disobedience. Consequently, one man can plunge a family into a life of tragedy. A person may not seem important to others in society, but he is very important to his family. Neither a man nor his family will achieve much in life if they continually sit on the shady side of the family tree. By one man, our world was thrown into chaos, but one man can make a difference in his family by being united with God and His principles. By one Man the human race was restored to God. When a man's ways and God's ways become as one, there exists unity which becomes evident in the family of God.

II. One Man's Obedience - Romans 5:15

Adam was the first man who disobeyed God. The penalty for that sin was death. He did not die a physical death that day, but he died a spiritual death. Jesus Christ, the Son of God, was one hundred percent man and one hundred percent God. He willingly obeyed God the Father by coming to earth and dying on a cross. He lived a sinless life and obeyed His Father in all points. By this one man's obedience, any person can be united with God by the act of regeneration by the Holy Spirit. Through the obedience of one Man, unity with God can be experienced by anyone who seeks God.

III. One Way to Heaven - Acts 16:31

The sin of man prevents him from entering Heaven. Any sin must be paid. No man can pay for his own sin because of the inherent sin nature received at birth. Christ Jesus, the perfect man, died on a cross to pay for the sin of the world, and only through believing on Him can anyone enter Heaven. He is the only way which a person can take to achieve this goal. No other man has lived a perfect, sinless life. He is the one way to Heaven. When a person admits he is a sinner, believes on the Lord Jesus Christ and calls upon His name, then that person becomes one in agreement with God. One is the number of unity.

CHAPTER 5
THE NUMBER TWO ~ "WITNESS"

The Source

The first mention of the number two in the Bible is in Genesis 1:16 where the two great lights are mentioned, the sun and the moon. The Mosaic law required two witnesses before a matter would be established (Deuteronomy 19:15). Amos the Prophet (Amos 3:3) asked, "Can two walk together, except they be agreed?" Christ condensed the entire Old Testament to two commandments and stated that the entire law and prophets hung on two commandments (Matthew 22:40).

The Symbolism

The digit two symbolizes "witness." As the Old Testament in Deuteronomy required two witnesses to establish an event, Jesus reaffirmed the requirement in John 8:17, "that the testimony of two men is true." We have two testaments or covenants as a witness to the world, the Old and New Testaments. There were two main components in the construction of the tabernacle, gold and shittum wood. These speak of the testimony of the two-fold nature of Jesus Christ, His divinity and His humanity. Jesus Christ is the second Person in the Trinity. He is the witness given to the world. There were two witnesses on the road to Emmaus. There will be two end-time witnesses upon this earth in the great tribulation (Revelation 11:3).

The Sermon

The Manner of Our Witness

I. The Walk of Our Witness - Romans 6:4

As a person walks with God and before this world, there is a witness given. The digit two symbolizes "witness." Many people watch a person and judge them by their actions. The believer is to walk in this world as if God were walking beside them. It has been said that actions speak louder than words and tell less lies. By walking with God each day, the believer gives a witness to the world. Peter's walk with God was so close that loved ones with illnesses were laid by their relatives and friends so that his shadow would fall upon them to be healed. Every believer has a shadow ministry. It is imperative that the believer walk with God to give the right testimony in his community.

II. The Talk of Our Witness - Philippians 1:27

When our conversation is about heavenly things, it is a witness by the believer to the world. Our talk should match our walk with God. Some people can talk Christianity by the yard, but they cannot or will not walk it by the inch. The walk with God and the talk is equally important. It has been estimated that there are more than 200,000 useless words in the English language and from some believers you hear all of them. Many people use a tub full of words to express a spoonful of thought. Jesus stated that the believer is to be clear and to give a "yes" or "no" answer. Our witness is greatly multiplied when our talk matches our walk for Christ.

III. The Balk of Our Witness - Romans 12:2

As our walk and talk are important, our balk is important. The things which we hesitate to accept, matter in this life. A Christian is a living sermon whether or not he preaches a word. A good example is often a good witness to the world when the believer refuses to accept a faulty concept or undertaking. Refusing to be engaged in a non-Biblical affair gives the onlookers a witness. The world's shortest sermon is found on a traffic sign, "Keep Right." The things at which a believer balks is a witness to the world.

CHAPTER 6

THE NUMBER THREE ~ "COMPLETE"

The Source

The first mention of the number three in the Bible is in Genesis 6:10. In this verse God gives the names of the three sons of Noah. In Genesis 15:4-9, God makes a covenant with Abraham that his descendants will be increased to the number of the stars in heaven through the child that will be born to him. In response to Abraham's question of how he would know this covenant was fulfilled, God told him to sacrifice three animals of the age of three. After the third day of travel, Abraham, Isaac, and the servants arrived at the foot of Mount Moriah where Abraham would offer Isaac to God (Genesis 22:4). The number three appears again in the Old Testament when the mother of Moses hid him for three months to prevent him from being killed by Pharaoh (Exodus 2:2).

In the New Testament the number three is plentiful. Christ was tempted in the wilderness and responded from the book of Deuteronomy three times in response to the temptation of the devil (Matthew 4). When questioned about His authority to perform miracles, Jesus told the Pharisees that the sign He would give them would be the sign of Jonah. As Jonah was in the belly of the whale three days, Christ stated that He would be in the heart of the earth three days and three nights. On the Mount of Transfiguration, Peter wanted to build three tabernacles (Matthew 17:4). Jesus prophesied his death, burial, and resurrection after three days of His death in Mark 8:31. Saul was without sight for three days in Acts 9:9.

The Symbolism

The number three symbolizes completeness. The numbers one, three and seven are used the most in the Bible. The number three is found everywhere in nature. The material world consists of space, matter and time. Space is described in length, breadth, and height. Matter is found in gas, liquid, or solid state. Time is described as past, present, or future. Even the actions of man are revealed in the number three with thought, word, and deed. The spiritual symbols of completeness are revealed in God who has three attributes, omniscient, omnipotent, and omnipresent. Christ's complete power over the dead is revealed in three ways: by raising the daughter of Jairus who had just died; by raising the son of a widow whose boy was on the way to the cemetery; and by raising Lazarus from the dead who had been in the grave for four days. As Jonah was in the belly of the whale for three days and three nights, Christ Himself would be in the heart of the earth for three days and three nights. There were three hours of darkness upon the earth which was the complete time needed while the holiness of God manifested itself with His Son being made sin for the believer. Christ arose on the third day which revealed God's satisfaction of His Son's sacrifice. The divine completeness of Christ's shepherd care is revealed by Him being the good Shepherd in His death, John 10:11; the great Shepherd in His resurrection, Hebrews 13:20; and the chief Shepherd at His coming, I Peter 5:4. The complete Godhead is revealed in Matthew 28:19 where the believers are commanded to go into all the world and baptize in the Name of the Father and of the Son and the Holy Ghost.

The banner of David or commonly called the "Star of David" is a picture of completeness. The Hebrew language has no vowels. The twenty-two letters of the Hebrew language have their vowel sounds built in the letters. David's name in English would be reduced to the three Hebrew letters, DALETH, VAU, DALETH. It would look like DVD. The symbolism is very powerful. In forming his banner for his military forces, David placed the first daleth which looked like a delta vector on the banner. Then he placed the second daleth inverted on the banner atop of the first daleth. He then placed the vau in the center. In the old Hebrew the VAU resembled a "T" or a cross. The banner of David was a hexagon or two equilateral triangles placed atop each other. The triangle pointing upward represents Christ in His fleshly body which was created in the image of God with body, soul and spirit. The triangle pointing downward represents God's image of the Father, Son and the Holy Spirit. The VAU or cross in the middle represents the cross upon which Jesus Christ died. The banner of David was symbolic of Jesus Christ who was 100% man and 100% God who died on a cross. The banner of David is a picture of completeness. Christ was completely man, completely God and the complete sacrifice for the sins of the world.

The Sermon

The Believer Is Complete

I. Complete in Salvation - I Thessalonians 4:16

When a person is born again, the soul and spirit are saved. When his body dies, then at the catching away of the church, the body, soul, and spirit are reunited. The believer is complete in salvation. Just as the soul and spirit are saved, one day the body will be saved and transformed.

II. Complete in Security - II Timothy 1:12

The believer in Christ is complete in the security of his decision to live and serve Christ. In II Timothy 1:12, Paul was persuaded that God was able to keep that which he had committed unto Him.

III. Complete in Service - Colossians 3:23

The believer in Christ is complete in service for the Lord Jesus Christ. A self-centered life will leave the believer not experiencing a fulfilled life. A life centered on serving Christ is a life of satisfaction.

CHAPTER 7

THE NUMBER FOUR ~ "THE WORLD - MANKIND"

The Source

The first mention of the number four in the Bible is found in Genesis 2:10 where a river parted as it went out of Eden into four heads. Similar to the journey of Adam and Eve, who represent all mankind, the water from the garden of Eden flowed into four parts to make its journey on the face of the earth. After the great flood, the location of these four rivers was probably radically changed.

On the fourth day of creation, the stars were created to provide the synchronization of the natural events upon all the earth (Genesis 1:18). Four world powers were revealed by Daniel: Babylonian, Medo-Persian, Grecian and Roman (Daniel 9:32-34). The four gospels contain the declaration of God's love for the world. According to Matthew, there are four kinds of soil in the world in which the Word of God is planted (Matthew 13). The church is given the responsibility to witness to four areas of the world: Jerusalem, Judea, Samaria, and the uttermost part of the earth (Acts 1:8). The fourth miracle in St. John's gospel is the miraculous feeding of the five thousand (John 6:1-13). There are four things listed in Revelation 21:4: death, sorrow, crying, and pain which plague mankind.

The number four is the first nonprime number which follows a prime number in the number system. Zero and one are not defined as prime numbers. The number four is the product of two times two. Two is the first prime number and the only even prime number. The number four has three factors: one, two, and four.

The Symbolism

The number four symbolizes the world or mankind as a whole. There are four corners of the earth (Revelation 7:1) and four directions on the earth: north, south, east and west. On the fourth day of creation, stars were created for four reasons: signs, seasons, days, and years (Genesis 1:18). The four divisions of the day are evening (sunset), midnight, early morning (cock crowing), and morning (noon) according to Mark 13:35. There are four seasons of the year. There are four lunar phases of the moon. There are four heavenly bodies: earth, sun, moon, and stars. The spiritual meaning of the number four is the world or mankind as a whole. In Genesis, the divisions of the world are given: lands, families, tongues, and nations (Genesis 10:5). The fourth book of the Old Testament, the book of Numbers, tells of the wilderness journey of the believer through the world. The dream of Nebuchadnezzar foretold the four great world powers (Daniel

2:31 - 36). Four Gospels contain the declaration of God's love for the world (John 3:16). There are four soils in which the Word of God can be placed (Matthew 13): the wayside, the stony ground, the thorny ground and the good ground. The fourfold witness of the church is in Jerusalem, Judea, Samaria, and the uttermost part of the earth (Acts 1:18). The fourth miracle in the book of St. John was the miracle of feeding five thousand. The great multitude seen in Revelation 7:9 tells of the four sources of man: nations, kindreds, people, and tongues. There are four sides to the new Jerusalem (Revelation 21:13). There are four worldly things missing in the new Jerusalem: death, sorrow, crying and pain (Revelation 21:4).

The Sermon

The Condition of Man

I. Mankind is Utterly Lost - Romans 3:10

The spiritual condition of man is one of sadness. Once living in the Garden of Eden, man is now living in a world with a curse. His final destiny without the influence of God is hell.

II. Mankind is Unilaterally Looking - II Corinthians 4:4

A glance at the religions of the world indicates that man is looking for a god to worship. The god of this world, the devil, has blinded the minds of people to look for a god other than the real God Jehovah. From the slopes of Nepal to the plains of India, man is searching for a god to worship.

III. Mankind is Uniquely Loved - John 3:16

Although man is utterly lost and looking for a Savior, the One and True God has provided the remedy for every person. Man has been uniquely loved by God in Heaven. The gospel is stated concisely in John 3:16. The major action in this verse is "love." God has demonstrated His love for man in an extreme manner: by giving His Son to die for the sins of man. Man is truly uniquely loved.

CHAPTER 8

THE NUMBER FIVE ~ "GRACE"

The Source

The first mention of the number five in the Bible is found in Genesis 18:28. The law of first mention clearly gives the interpretation of this number. The law of first mention is this: The first time a word, incident or phrase occurs in the Bible it gives the key to its meaning elsewhere in Scripture. The first mention of the number five in the New Testament is in Matthew 14:17 where five loves and two fishes are given to the Savior by a small lad. David selected five smooth stones to battle Goliath in I Samuel 17:40. Mephibosheth was five years old when he was crippled in II Samuel 4:4. In II Chronicles 3:11, the wings of the cherub over the mercy seat were five cubits. Five porches were at the pool of Bethesda in John 5:2 where the impotent man lay.

The Symbolism

The idea associated with the number five in the scriptures is the grace of God. In seeking the forgiveness of God for the residents of Sodom, Abraham used the number five to determine the extent God would demonstrate His grace (Genesis 18:28). The anointing oil for the holy tabernacle and vessels had five ingredients (Exodus 30:23-25). God uses His hand of five fingers as a symbol of grace throughout the Bible according to Psalm 40:1-3, Luke 1:66 and Acts 11:21. The number five is symbolic of the grace of God.

The Sermon

The Grace of God

I. The Definition of Grace - Titus 3:4.

The grace of God is defined in Titus 3:4 where it says, "But after that the kindness and love of God our Saviour toward man appeared." Grace is the undeserved merit or love of God towards man.

II. The Demands of Grace - Titus 2:12.

The grace of God teaches that the believer should deny the world and its lusts, and live soberly, righteously, and godly in this present world. The believer should live with an expectation of the Lord Jesus returning to earth for the church at any time.

III. The Depths of Grace - Romans 8:28

The grace of God can reach lower and higher than any person could ever dream. Even in the worse of times, Paul told the Romans, "And we know that all things work together for good to them that love God, to them who are the called according to His purpose (Romans 8:28).

CHAPTER 9

THE NUMBER SIX ~ "MAN"

The Source

The first mention of six in the Bible is in Genesis 1:31 concerning the sixth day of creation. God set up six cities of refuge in Joshua 20:7-8 to which the unintentional slayer of a man could flee to escape judgment. Goliath of Gath had a height of six cubits. He wore six pieces of armor (First Samuel 17:4-7). In Daniel 3:1, Nebuchadnezzar erected an image which was sixty cubits high and six cubits broad. In the ten commandments, the sixth commandment chronicles the terrible sin of man against man which is murder (Exodus 20:13). In Proverbs 6:16, there are six things which God hates in man. There were six water pots in John 2:6 when Jesus performed His first public miracle of changing water into wine. Six times the charge of demon possession was made against Jesus (Mark 3:22; Luke 11:15; John 7:20; John 8:48; John 8:52; and John 10:20). Six persons united to testify that Christ was innocent: Pilate - Luke 23:14; Herod - Luke 23:15; Judas - Matthew 27:4; Pilate's wife - Matthew 27:19; the dying thief - Luke 23:41, and the Roman centurion - Luke 23:47. On the sixth hour of the day, darkness covered the land as Christ hung on the cross (Matthew 27:45).

The Symbolism

The idea associated with the number six in the Holy Scriptures is man. It was on the sixth day of creation that man was created. Man, who has the nature of Adam, had six cities to which to flee when mistakes were made (Joshua 20:7-8). Goliath of Gath, the champion of the Philistines, had a height of six cubits and wore six pieces of armor as specified by I Samuel 17:4-7. Goliath represents man by his nature. He fought against God. The image erected by Nebuchadnezzar was sixty cubits high and six cubits broad according to Daniel 3:1. The statue represented the four kingdoms of man that God had prophesied would come. The water pots, which Jesus instructed the servants to fill with water, represented man's failure to bring blessings. In I Kings 10:14, six hundred sixty-six talents of gold were brought to Solomon in one year, but even this amount of wealth could not bring peace to the kingdom. In Ezra 2:13, the children of Adonikam who returned from the captivity numbered six hundred sixty-six. The name, Adonikam, means *the Lord of the enemy.*

There are three men who stand out in scripture as the avowed enemies of God. Goliath was six cubits high. Nebuchadnezzar set up an image which was sixty cubits high and six cubits broad, and the Antichrist in the book of

Revelation has the number six hundred sixty-six. Throughout the Bible, the number six is associated with the idea of man.

The Sermon

The Means of Atonement

I. The Nurture of Almighty - John 3:3

Man was created in the image of God and he was nurtured by Him. Man is lacking in the spiritual realm and needs to be born again. The first birth is not sufficient. The spiritual nature of man is lacking. He needs the nurture of God.

II. The Nature of Adam - Romans 5:12

Man is a fallen creature of God. Through the entrance of sin in the life of one man, Adam, all men became sinners. No one born into this world is exempt from this problem except Jesus Christ who was born of God. Man is a sinner.

III. The Need of Atonement - Romans 5:11

To make the transition from sinner to saint, man needs an atonement. He needs a covering for his sin. Christ is the atonement who died on a cross to pay for the sins of the world.

CHAPTER 10
THE NUMBER SEVEN ~ "PERFECT"

The Source

Seven is the one of the most used numbers in the Bible. There are seven notes on the musical scale. There are seven primary colors of the rainbow. Life functions in cycles of sevens. Cell changes are made every seven years in the human body. There are seven bones in the neck; seven bones in the face; seven bones in the ankle, and seven holes in the skull. The life of civilization is centered on the week or seven days. Children are usually born in 280 days which is a multiple of seven. God created the world in six days and rested on the seventh day (Genesis 2:2). There are seven feasts of Jehovah in the Old Testament. Jericho was conquered by the children of Israel marching around the city in seven consecutive days. Christ spoke seven sayings while on the cross of Calvary. Seven men were chosen as deacons of the first church. Seven letters were written to seven churches by the Apostle John on the Isle of Patmos.

The Symbolism

The number seven in the Bible is a symbol of perfection or maturity. God rested on the seventh day in the knowledge of creation's perfection. Enoch, the seventh from Adam was not, for God took him (Genesis 5:24).

Moses was the seventh from Abraham and the perfect person to lead the children of Israel out of Egypt. Abraham was given a sevenfold blessing by God (Genesis 12:2-3). The nation of Israel was given a sevenfold promise by God in Exodus 6:6-8. There were seven feasts of Jehovah of which some lasted seven days. In conquering Jericho, God instructed Joshua to have seven priests carrying seven trumpets to march around the city. On the seventh day, the priests preceding the Ark of the Covenant, compassed the city seven times instead of the usual one trip around the city. Matthew 12:45 speaks of seven evil spirits. Mary Magdalene had seven demons possessing her life. Seven men were chosen as deacons in the early church in Acts 6:3. Seven epistles were written to the churches giving perfect instruction in all matters pertaining to life and godliness. In the book of Revelation, seven letters were written to the churches of Asia and give the perfect, inspired history of the church. In the book of Revelation seven of the following is given: candlesticks, stars, lamps, angels, spirits, seals and plagues.

The number of the Bible is seven. The Apostle Paul in II Timothy 3:16 gave to his student, Timothy, the purpose of the

Bible. It is for doctrine, for reproof, for correction, and for instruction. Bible doctrine tells us what is right. Reproof tells us what is not right. Correction tells us how to get right, and instruction tells us how to stay right. The Bible is divided into two sections, The Old Testament and The New Testament. The number of the Old Testament is four which is the number of mankind. This portion of our Bible tells us of the fall of man and is centered on a four word question, "Where is the lamb?"(Genesis 22:7). The number of the New Testament is three which is the number of completeness. God's plan of salvation is made complete in the New Testament by the first coming of Jesus as the sacrifice for the world. The New Testament is centered on the three word statement of John the Baptist concerning Jesus, "Behold the Lamb" in John 1:29. Four plus three is seven and seven is the number of perfection. Mankind, four, plus the completeness, three, from God, equals a person, seven, who is perfect and mature in the eyes of God, The Bible is the only means to reach this stage of perfection.

The menorah or the golden candlestick in the tabernacle and the temple is associated with the number seven. It represents the perfect light which comes from the Holy Spirit symbolized by the olive oil. The menorah was in the shape of seven lamps with the center lamp higher than the other six. God gave Moses the instructions for the menorah. The middle lamp was called the servant lamp. Each lamp held "six eggs" measure of olive oil which would last one day. The temple built by Solomon had ten menorahs (I Kings 7:49). After the death of Christ on the cross, the servant lamp refused to burn from 30 A.D. to 70 A.D. at which time the temple built by Herod was destroyed by Roman soldiers. History records that the menorah was carried to Rome.

The first seven books of the Bible, Genesis through Judges, form a menorah. The fourth book of the seven set books is Numbers. In Numbers 8:2-4, Aaron lit the menorah. There were seven days of creation and rest. On the fourth day, God created the sun, the servant lamp. Josephus reported that the design of the menorah in the tabernacle represented the sun and the planets. There were seven known heavenly bodies in the time of the Old Testament: Mercury, Venus, Earth, Mars, Jupiter, Saturn, and the Sun.

Assigning the first seven numbers in the number system (0, 1, 2, 3, 4, 5, and 6) to the seven lamps of the menorah, places the number three on the middle candle. Three symbolizes completeness and Christ was complete in providing the light to the world of the gospel. He is the light.

The Sermon

The Perfection in the Life of the Believer

I. The Perfection of Our Redeemer - I Peter 1:19

Christ was perfect in His conception. He came to earth as the Redeemer of the world. The virgin Mary was chosen by God to receive His body. Mary provided nothing to His makeup. She did not provide His blood. His blood was from His Father in Heaven. She simply provided food from her womb, food from her breasts, and food from her table. He was perfect for He was and is the Son of God. He was perfect in His childhood, in His ministry, and in His death, burial and resurrection. He was the Lamb of God who was without blemish and without spot. He was the perfect sacrifice for the sins of the world.

II. The Perfection of Our Redemption - Romans 3:24

Redemption of a person is only as perfect as the Savior in whom one believes. Since Jesus Christ is the only sinless person from birth to death, He is the only perfect sacrifice. The redemption of the believer is safe and secure through Christ.

III. The Perfection of Our Reward - I Corinthians 2:9

Neither the eye, the ear, nor the imagination can conceive the reward for the believer of Christ. The reward will be perfect according to the gift of God to the believers.

CHAPTER 11

THE NUMBER EIGHT ~ "A NEW BEGINNING"

The Source

The first mention of the number eight is found in Genesis 17:12. The Law of First Mention is one of the many rules which provide valuable information in interpreting a portion or word of Scripture. The law is simple: The first time a word, incident, or phrase occurs in the Holy Scriptures, it gives the key to the meaning when the word is used elsewhere in the Bible. Also the Law of Simple Association of Ideas was proposed by Palmoni who said, "However we may explain it, certain numerals in the Scriptures occur so often in connection with certain classes of ideas, that we are naturally led to associate the one with the other."

God stated to Abraham that the male children should be circumcised when eight days old (Genesis 17:12). In Genesis 7:13, eight people went into the ark to escape the judgment of the flood. Noah was the eighth person from Adam when the earth was covered with water (II Peter 2:5). The priests, after being consecrated for seven days, entered upon their work on the eighth day (Leviticus 8:33). In Luke 9:26-29, Jesus speaks of the Kingdom of God, and eight days later He demonstrates for Peter, James, and John the Kingdom of God by transforming Himself into His glorified body. On the first day of the week, the women found the tomb of Jesus empty. In John 20:26, Jesus appeared to the disciples after eight days.

The Symbolism

The number eight is the symbol of a new beginning. It is associated with resurrection and regeneration. It shows the beginning of a new order of things. The male child who was circumcised on the eighth day entered into the covenant with God as in Genesis 17:9-14. The new birth of the believer is also associated with this operation in Colossians 2:11. As the foreskin of the child is removed and discarded, the believer puts off the body of the sins of the flesh by the circumcision of Christ. The discarding of the sins starts the new life for the believer. The eight people who entered the ark to escape the flood went into it for a new beginning upon the earth. There was a new world and a new beginning for Noah and his family. A new work started on the eighth day for the priest after being consecrated for seven days in Leviticus 8:33. Psalm 8 describes the new kingdom of Christ. How excellent is Thy name in all the earth! Christ arose on the first day of the Jewish week and appeared to the disciples. Eight is the number of a new beginning.

The Sermon

The New Ways of the Believer

I. The New Life of the Believer - John 3:3.

Jesus used the term "born again" with Nicodemus in a private setting. The believer has a new life in Christ at the moment of salvation. The new life is one centered in Christ instead of in self. The believer of Jesus Christ has a new beginning in his life. He has a new spiritual nature and a new life to live.

II. The New Love of the Believer - II Corinthians 5:17.

The apostle Paul tells the Corinthians "Old things have passed away, all things are become new" (II Corinthians 5:17). The old love for the world passes away when a person is "born again" The new love is the new law for the Christian.

III. The New Language of the Believer - Mark 16:17.

The new tongue mentioned in Mark 16:17 relates to the new language of the believer. The old language of the world is no longer appropriate for the believer with a new spiritual walk.

CHAPTER 12
THE NUMBER NINE ~ "DIVINE ORDER"

The Source

The first mention of the number nine is found in Leviticus 23:32 which was the start of the day of atonement. On the fifth day of the feast of tabernacles, nine bullocks were offered, along with two rams and fourteen lambs (Numbers 29:26). In Haggai 1:11 there are nine judgments given against the children of Israel for neglecting the house of God. In the New Testament the number nine (Luke 17:17) is pinpointed when Jesus asks, "Where are the nine?" In Mark 15:34, the translation of the words of Jesus upon the cross, "My God, My God, why hast thou forsaken me?" are nine English words. From the Sermon on the Mount (Matthew 5:3-12), there are nine beatitudes. The ninth hour was a traditional time of prayer (Acts 3:1; Acts 10:30). The nine gifts of the Spirit are given in I Corinthians 12:8-10. The fruit of the Spirit (Galatians 5:22-23) is revealed as being ninefold.

The Symbolism

The idea or symbolism associated with the number nine is divine order or divine completeness. The number eight represents a new beginning or the new birth. The number nine is a heavenly order instead of an earthly order. After the new birth, the fruit of the Spirit is expected from the good tree. There is a divine order for spiritual things. The day of atonement (Leviticus 23:32) began at the evening of the ninth day of the month and continued until the evening of the tenth day. The day of atonement was the day of divine completeness for the payment of sins. The fruit of the Spirit in Galatians 5:22-23 reveals this. The nine beatitudes given by Jesus are given in a divine order. The nine gifts of the Spirit comprise a divine order and a divine completeness of the work of the Holy Spirit. In Exodus 27:12, the Tabernacle was fifty cubits per length, and there were ten pillars on the long sides which made nine spaces between the pillars. The Holy Spirit came on the day of Pentecost fifty days after Christ arose. The nine spaces in the wall hangings, along with the dimensions of each side of the tabernacle of fifty cubits, connect the number nine with the work of the Spirit and shows the fruit of the Spirit disguised in the Old Testament.

The Sermon

The Divine Order

I. The Divine Order of Salvation - Acts 8:37

The new birth must occur before the fruit of the Spirit can develop in a person's life. The divine order of salvation must be followed. The eunuch was first saved and then baptized. The birth must be first, then the growth or fruit of the believer will occur. A person's desire to reform first, then be saved may be in vain. Without the aid of the Holy Spirit, any reformation will be superficial. A lot of people use mighty thin thread when mending their ways.

II. The Divine Order of Sanctification - John 17:17

Salvation must come before a person can be separated from the things of the world and set apart to do the work of God. The separation from the world cannot occur before God places His Word and His work in the heart of the believer.

III. The Divine Order of Service - Mark 4:20

Christ spoke of three quantities of fruit bearing: thirtyfold, sixtyfold, and a hundredfold (Mark 4:20). Christ spoke of three levels of fruit bearing: fruit, more fruit, and much fruit (John 15:1-8). Paul spoke of three wills of the believer; good, acceptable, and perfect will (Romans 12:2). There is a divine order a believer follows in his service for God.

CHAPTER 13

THE PRIME NUMBERS

The Definition and Significance of a Prime Number

To fully comprehend the divine design of digits, some terms need to be defined and understood. A prime number is a number that can only be divided by itself and one. The prime numbers are the building blocks of the number system. The Greek geometrician and educator in Alexandria, Euclid, gave proof there is no limit to the number of prime numbers. The number zero is not a prime number. The number one is not a prime number. It is often called the unit. Two is the first and only even prime number. The remainder of the prime numbers would be odd numbers. The first eleven prime numbers are 2, 3, 5, 7, 11, 13, 17, 19, 23, 29, and 31. The only divisors of these numbers are the number itself and the number one.

The Interdependence of the Symbolism of the Prime Numbers and the Nonprime Numbers

The symbolism of the prime numbers reveals God's plan for man. Two is symbolic of witness. Three is symbolic of complete. Five is symbolic of grace. Seven is symbolic of perfection. Eleven as we will find is symbolic of judgment. The symbolism of these prime numbers describe the sequence of events in the life of a believer. First, a person must hear the

Word of God. His soul must be in unity (one) with the hearing of God's Word. The witness (two) of the Word of God brings the person to a decision. The believer is made complete (three) when the person surrenders to Christ and is born again. This occurs when the grace (five) of God enters the person's life. At the end of this life and entering into the presence of God, the believer's viewpoint is one of perfection (seven). Judgment (eleven) of the believer at the Judgment Seat of Christ is the next step.

The first four nonprime numbers are four, six, eight, and ten. Four is the number of the world. Six is the number of man. Eight is the number of a new beginning. Ten is the number of order. These four numbers are found symbolically in one of the most concise verses of the Bible. John 3:16 says. "For God so loved the world (four), that He gave His only begotten Son, that whosoever (six) believeth in Him should not perish (eight) but have everlasting life (ten).

CHAPTER 14
THE NUMBER TEN ~ "ORDER"

The Source

The first mention of the number ten in the Bible is in Genesis 16:3. In Genesis chapter one, the words, "God said" occur ten times. There were ten plagues upon Egypt (Exodus 7:12). Ten commandments were given to Moses from God (Exodus 20). The tithe is one out of ten (Leviticus 27:32). The image of Nebuchadnezzar had ten toes which represented ten kingdoms (Daniel 2:42). The Passover lamb was taken on the tenth day (Exodus 12:3). The parables of the kingdom are ten in number in Matthew's gospel; seven in chapter thirteen and three in chapters twenty-two and twenty-five. There were ten pieces of silver in Luke 15:8 and one was missing. Romans 8:38-39 describes the tenfold security of the child of God through faith. Ten servants were entrusted with ten pounds and one was rewarded with authority over ten cities (Luke 19:13-17). Matthew 25:1-13 tells the parable of the ten virgins.

The Symbolism

Where the number nine speaks of the order in the divine realm, the number ten tells of the order upon earth. To eventually establish a family in the promised land after ten years, Sara persuaded Abram to take Hagar and have children by her to provide an heir or provide order in their family (Genesis 16:3). The ten plagues upon Egypt established order, in that slavery was wrong and all persons should be free (Exodus 7:12). The ten commandments were the beginning of many laws from God to establish order upon the earth not only for the children of Israel, but for all man (Exodus 20). The giving of the tithe established the order of giving for man to God (Leviticus 27:32). The ten pieces of silver mentioned in Luke 15:8 were probably the dowry of the woman who lost one of them. This was probably her inheritance which was established to provide order in her society in the times of Jesus. The parable of the ten virgins reflects that ten persons was the legal number necessary for a Jewish function, or wedding in this case, to be in order. Ten symbolizes order on the earth. It is associated with the civil and religious laws. Laws of the land are to provide order for the society which enacted the laws.

The Sermon

The Laws of God.

I. God's Law of Lessons - Leviticus 27:34

The Mosaic Law is the foundation of civility in the world. The book of Leviticus provided a clear order in the lives of the Jewish people. Almost every step of their lives was directed by the commands given by God to Moses. Even today, many Jewish people fashion their lives around the book of Leviticus.

The book of Leviticus first directs men in the proper manner to worship God with sacrificial offerings. Leviticus deals with social problems such as leprosy and other unclean issues. The lessons given in the Book of Leviticus not only regulated man's conduct toward another person, but towards God and themselves.

II. God's Law of Life - Joshua 24:14 - 15

A person without order in their life is digressing toward trouble in this life. Joshua gave a "state of the union" address to the children of Israel in Joshua 24:14-15. He gave them two choices: obey God and be blessed, or disobey God and experience the curse. God's Law of Life is to obey Him and experience His blessings in this life and the world to come.

III. God's Law of Love

The Apostle Paul stated that to fulfill the Mosaic Law, the believers were to love one another (Romans 13:8). This was a new law in that the Jewish person was taught from a child that the remedy for social injustice was "an eye for an eye," The new law was one of love. It is the royal law (James 2:8). Jesus summarized the commands of the entire Old Testament into two commands.

CHAPTER 15

THE NUMBER ELEVEN ~ "JUDGMENT"

The Source

The first mention of the number eleven is in Genesis 32:22 relates to the meeting of Jacob and Esau. Canaan, the grandson of Noah, had eleven sons (Genesis 10:15-18). Zedekiah reigned eleven years in Jerusalem (Jeremiah 52:1-2). An eleven day journey brought the children of Israel to Kadesh-barnea (Deuteronomy 1:2). Eleven disciples were remaining when Judas left the inner circle of followers of Jesus to betray Him (John 13:30).

The Symbolism

The symbolism of the number eleven is judgment. The number ten represents order which is following the law and being responsible. Going one step beyond order brings judgment and disorder. Around the first mention of the number eleven in the Bible is the idea of judgment. Jacob wrestled with God in Genesis 32:24. Jacob also wrestled with facing his brother Esau as Judgment Day had arrived in his life. Noah pronounced judgment upon Canaan in Genesis 9:20-25, and in Genesis 10:15-18, it is revealed that Canaan had eleven sons who would also suffer because of this judgment. Counting the ten plagues in the land of Egypt and the parting of the Red Sea and destruction of the army of Pharaoh, there were eleven judgments upon Egypt. At Kadesh-Barnea, the children of Israel brought judgment upon themselves by refusing to possess the promised land (Deuteronomy 1:2). Zedeiah was an evil king in Jerusalem (Jeremiah 52:1-2) and after eleven years of reigning, he was captured and taken to Babylon where judgment was executed upon him (Jeremiah 52:7-11). The number eleven is associated with judgment.

The Sermon

The Great White Throne Judgment

I. The One on the Throne ~ Revelation 20:11

The first impression of the Apostle John of the Great White Throne Judgment was not the throne itself, but the Person on the throne. John saw the power and authority of this Person for the earth and the heaven fled away and there was found no place for them (Revelation 20:11). The Person on the throne is revealed in the Scriptures as Jesus, the Son of God (John 5:22), for the Father will judge no person, but has given His Son the authority to judge all persons.

II. The Ones Before the Throne ~ Revelation 20:12

The ones before the throne are sinners. Not only will the earth give up its dead, but the seas (Revelation 20:13) will give up their dead. Death and hell delivered up their occupants and they stood before the Great White Throne. Small and great were the sinners. There will be no respect of persons at this Great White Throne Judgment. There will be no suspended sentences at this judgment.

III. The Ones Before God ~ Revelation 20:13-15

The Scriptures ensure that the judgment is conducted with fairness. There are two sets of books which will be used in this judgment. Each person alive has a book of their life. It is a book that records every word and every deed (Matthew 12:36) of each person on the earth. Another book is mentioned, and it is the Book of Life (Revelation 20:15). A final check is made in the Lamb's Book of Life to demonstrate to the sinner that there was a place for them in the Lamb's Book of Life, but they neglected to repent and be born again. Supreme justice will take place at the Great White Throne Judgement.

CHAPTER 16

THE NUMBER TWELVE ~ "DIVINE AUTHORITY"

The Source

The first mention of the number twelve is in Genesis 14:4 where a union of kings served Chedorlaomer, who was head of the union of kings. Ishmael begat twelve princes (Genesis 17:20). There were twelve stones in the high priest's breastplate (Exodus 28:17-21). The twelve tribes of Israel had twelve princes over them (Numbers 1:5-16). Twelve spies were sent to inspect the promised land (Numbers 13:1-133). Solomon had twelve cabinet members in his government (I Kings 4:7). There were twelve judges in the book of Judges. Jesus visited the temple at the age of twelve. Jesus chose twelve disciples. Jesus predicted a day when He, the Son of Man, would sit in the throne of His glory and promised that His disciples would sit upon twelve thrones judging the twelve tribes of Israel (Matthew 19:28). At His betrayal in the garden of Gethsemane, Christ rebukes Peter and tells him that He could command twelve legions of angels to deliver Him (Matthew 26:53). The number twelve appears in the Book of Revelation several times: (1) the woman with a crown of twelve stars (Revelation 12:1); (2) the twelve gates of New Jerusalem (Revelation 21:12); (3) at the gates, twelve angels

(Revelation 12:12); (4) the twelve foundations and in them the twelve names of the apostles (Revelation 21:14); (5) the tree of life bears twelve manner of fruits (Revelation 22:2); (6) New Jerusalem lies four square and measures twelve thousand furlongs (Revelation 21:16); and (7) the height of the wall is 144 cubits which is twelve times twelve (Revelation 21:17).

The Symbolism

The idea associated with the number twelve in the Bible is divine authority. This is not earthly or governmental rule, rather it is a heavenly rule or governmental perfection. The number eleven represents judgement and disorder. Twelve is the next step in correcting disorder by divine authority. Our calendar is divided into twelve months which reflects divine authority. The two sons of Abraham, Ishmael and Isaac, had the number twelve in their lives. Ishmael had twelve sons (Genesis 17:20) who governed the his affairs. Isaac had twelve grandsons who ruled over the twelve tribes of Jacob (Numbers 1:5-16). According to First Kings 4:7, Solomon had twelve cabinet members in his government to whom he delegated authority to rule over his kingdom. To govern the children of Israel between the times of the great leaders, Moses and Joshua, and the great kings, God placed twelve judges to provide divine authority for the children of Israel. At the age of twelve, Jesus went to

Jerusalem (Luke 2:42) and was found by his mother and foster father in the midst of the Mosaic lawyers discussing the matters of the law. His divine authority was revealed at the age of twelve where He astonished the religious leaders at His understanding and answers. To manage His spiritual kingdom, Jesus ordained twelve disciples (Mark 3:14). Matthew 19:28 discloses His future kingdom and some of the features with twelve thrones. Throughout the book of Revelation, the number twelve indicates the divine authority of God in governing His affairs. We see God's divine authority in the heavens with the two sets of twelve hours in a day, and the twelve signs of the heavens, the Zodiac, with 360 degrees of measurement.

The Sermon

Divine Authority

I. The Principle of Delegation ~ I Kings 4:7

Solomon knew the advantage of not only having a multitude of counselors in his administration but having cabinet members who would oversee the functions of his kingdom. Solomon had twelve cabinet members (First Kings 4:7) to whom he delegated his authority to carry forth his policies. Moses was advised by his father-in-law, Jethro, to delegate the judgments needed in the lives of the Jewish people (Exodus 18:19-22). God has used this principle to multiply His work.

Jesus had twelve disciples and not only was this a means of teaching and equipping them for the work of God ahead, but also to delegate to them the responsibility of spreading the gospel of Christ to the world.

II. The Principle of Dedication ~ I Corinthians 15:58

Along with the principle of delegation is the principle of dedication. The person with authority to do the will of God must be a person of integrity. A believer should be humble when considering that they were chosen by God to work in any capacity in His kingdom. Honoring God with humbleness is a rare commodity. Some believers are proud that they are humble. The principle of dedication amplifies the idea of delegation in the work of God.

III. The Principle of Decrease ~ John 3:30

John the Baptist delegated the preaching of the gospel to his disciples. The principle of decrease was found in John the Baptist at his apex. No greater prophet was there than John the Baptist, yet he was willing to decrease and let Christ have the preeminence. If the person to whom the authority has been delegated does not assume the principle of decrease, then personal goals and aims may interfere with the work of God. The principle of decrease sharpens and improves the principle of delegation.

CHAPTER 17

THE NUMBER THIRTEEN ~ "REBELLION"

The Source

The first mention of the number thirteen is in Genesis 14:4. Ishmael, the son of Abraham, was circumcised at the age of thirteen (Genesis 17:25). Solomon took thirteen years in building his house (First Kings 7:1). In Esther 3:8-13, Haman, the enemy of the Jews, had a decree signed on the thirteenth day of the first month to have all the Jews put to death on the thirteenth day of the twelfth month. In the heavenly company of Jesus and His disciples, there were thirteen people. Judas Iscariot is the thirteenth person in this list. In Mark 7:21-22, Jesus mentions thirteen traits of the rebellious heart. The word "dragon" which is used to describe Satan is mentioned thirteen times in the book of Revelation.

The Symbolism

In Genesis 14:1-4, Chedorlaomer, King of Elam, formed a union of kingdoms which conquered five adjoining kingdoms. In the thirteen year of this conquest, there was rebellion in the kingdom. Ishmael at the age of thirteen (Genesis 17:25), was circumcised only in the flesh and not in the heart. Rebellion has its seat in the heart of a person. Haman rebelled against the blessings of God through the Jewish people by attempting to annihilate them (Esther 3:8-13). The date of the decree and the date of the liquidation was on the thirteenth of the first and twelfth months. The number thirteen has often been assigned as an unlucky number. Some large hotels do not list a thirteenth floor but skip from the twelfth floor to the fourteenth floor. Some smaller airliners do not have a thirteenth row of seats. This has been associated with the listing of Jesus and His twelve disciples because Judas is the thirteenth person in the list (Matthew 10:1-4). Judas was a person full of rebellion. In Galatians 5:19-21, God lists the works of the flesh. The thirteenth in the list is heresies, which is profoundly at odds against God which is rebellion. In the list of twenty-three things of evil in humanity (Romans 1:29-31), the thirteenth is "haters of God" which is a clear sign of rebellion against God.

The Sermon

The Center of the Rebellion Against God

I. Rebellion by Violation - Mark 7:20

When His disciples ate bread without washing their hands and were accused of being unclean, it gave Jesus an opportunity to teach an important spiritual truth. The Pharisees immediately took opportunity to condemn the disciples of Christ and thereby condemn Christ Himself. This action did not take Jesus by surprise, instead, He took occasion to expound the true Word of God. He taught that rebellion against God was not neglecting to simply wash the hands, but it came from the violation of the person's heart by evil thoughts, which would lead to evil acts. Rebellion against God is first conceived in the heart of a person by vile meditations. Afterwards, the hands and feet rush to carry out this rebellion. Jesus emphasized to the Pharisees that it was not the outward act of neglecting to wash the hands, but the inward pollution of evil thoughts.

II. Rebellion by Violence ~ Mark 7:21 - 22

The violation of a person's heart with evil thoughts leads to rebellion against God by violence. Jesus stated that from the heart proceed, "evil thoughts, murders, adulteries, fornications, thefts, covetousness, wickedness, deceit, lasciviousness" (Mark 7:21-22). These are all acts of violence against humanity.

III. Rebellion by Viciousness ~ Mark 7:22.

Violation of another person is evil, but viciousness against another person is extreme wickedness. At the end of the list of evil actions, Jesus listed lasciviousness, an evil eye, blasphemy, pride, and foolishness. Rebellion against God eventually is viciousness against a person's neighbor. Christ gave the definition of the word "neighbor" as being someone with whom another person has contact. Lasciviousness, or an overt, offensive sexual desire towards another person is a vicious act.

CHAPTER 18

THE NUMBER FOURTEEN - "SALVATION"

The Source

The first mention of the number fourteen is in Genesis 31:41 when Jacob pronounced his accusations against Laban and stated that he had labored fourteen years for Laban's two daughters as his wives. Fourteen is the seventh nonprime number. In Exodus 12:6-7, the Passover lamb was slain on the fourteenth day. King Solomon completed on the fourteenth day the dedication of the temple (First Kings 8:65). In Ezekiel 43:17, the prophet Ezekiel described the area of the altar as a square of fourteen cubits. In Matthew 1:17, the number fourteen relates to the number of generations to the first coming of Jesus Christ. Paul encountered a storm in the fourteenth night (Acts 27:27) of the journey. In Galatians 2:1, Paul stated that fourteen years, after the revelation of Jesus to him as God, he journeyed to Jerusalem.

The Symbolism

Fourteen symbolizes salvation. This can be an earthly deliverance or a heavenly one. In Genesis 31:41, Jacob served his father-in-law, Laban, fourteen years and was delivered from his firm, authoritative grip. On the fourteenth day, the Passover lamb (Exodus 12:6-7) was slain to provide salvation to the sacrificial givers. The Temple built by Solomon was a great symbol of salvation. It was dedicated to God on the fourteenth day of the month (First Kings 8:65). The prophet Ezekiel clearly described the area of the altar in Ezekiel 43:17 which was associated with salvation. In Matthew 1:17, the number fourteen is stated, accompanying the first coming of Jesus Christ as the Savior of the world. The saving of the apostle Paul from the storm commenced on the fourteenth night of the journey (Acts 27:27). Fourteen years after Paul had been saved and had gone to Jerusalem, he returned (Galatians 2:1) with Barnabas and Titus to Jerusalem to settle the issue of the origin of salvation. Is it by works or grace? Fourteen is a symbol of salvation and three is a symbol of completeness. Israel was delivered from the plague in Egypt on the fourteenth day, and three days later they passed through the Red Sea. Their salvation was made complete, and they sang a song (Exodus 15:1-2) which said, "The Lord is my strength and song, and He is become my salvation."

The Sermon

The Salvation of the Believer

I. The Day of Deliverance ~ Exodus 14:13-14

God had planned the day of deliverance for the children of Israel. The Israelis had been living in Egypt for four hundred and thirty years, and the day of salvation was being revealed to them. God told Moses to tell the children of Israel that they would see the salvation of the Lord (Exodus 14:13). The entire events of deliverance for the children of Israel from Egypt was initiated by the offering of the Passover Lamb in Exodus 12:6. The day of salvation and deliverance came for the children of Israel as they escaped the Egyptians by crossing the Red Sea.

II. The Day of Duty ~ Exodus 12:41-42

The day of salvation was a day of duty. There were strict and precise directions given by Moses to the children for them to secure their salvation from the Egyptians. The Passover meal had exact regulations. It was the duty of the children of Israel to follow these directives to be saved. The children of Israel were given direct orders in how to leave Egypt and cross the Red Sea. The day of salvation brings the day of duty in the life of the believer.

III. The Day of Delight ~ Exodus 15:20

With the destruction of the enemy by the closure of the Red Sea, Miriam and the women of Israel rejoiced before the Lord Almighty by playing their timbrels and dancing. There was delight in the camp of the Israelis for their salvation and deliverance from the mighty forces of evil, the Egyptians. The salvation of the believer is truly a day of delight and rejoicing.

CHAPTER 19

THE NUMBER FIFTEEN - "REST"

The Source

The first time the number fifteen is used in scriptures is in Genesis 7:20 where the height of the water in the great flood was reached at fifteen cubits above any land feature. On the fifteenth day of the month, the children of Israel were given a social command in Leviticus 23:6-7. The number fifteen is found in Leviticus 23:34-35. King Hezekiah received a cure from his disease and given an additional fifteen years to his life (II Kings 20:6). In the conflict in the Persian empire, the Jewish people received a reprieve on the fifteenth day of the month in Esther 9:20-22. The small book of Ruth reveals that the fifteenth time the name of Naomi is used is in Ruth 3:1 where Naomi has rest. Bethany, where Lazarus was raised from the dead, and from whence the Lord ascended, was fifteen furlongs from Jerusalem (John 11:18). Paul's ship was anchored in fifteen fathoms on the fourteenth day after thirteen days of toil and trial (Acts 27:21). After three years in the Arabian desert, the apostle Paul went to Jerusalem to confer with the apostle Peter for fifteen days (Galatians 1:18).

The Symbolism

Clearly the idea associated with the number fifteen is rest. The Law of First Mention holds particularly true in this instance in Genesis 7:20 where the word fifteen is first found in the Holy Scriptures. After entering the ark, Noah and his family heard it rain for forty days and forty nights. The ark was lifted and the people outside the ark died. The highest mountains were covered by at least fifteen cubits of water. No one could survive outside the ark built by Noah. The waters rested when the height of fifteen cubits was reached by the water over the highest mountains. The order seems to be rebellion (13), then salvation (14), and then rest (15). The people of the earth rebelled against God. God provided a means of salvation which was the ark. Those who entered the ark found rest.

The order of fifteen following fourteen is found also in Leviticus 23:4-8. On the fourteenth day of the first month is the Lord's Passover (salvation). On the fifteenth day of the month is the feast of the unleavened bread where for seven (perfection) days the children of Israel must not eat any leavened bread. The fifteenth (rest) day of the month followed the fourteenth (salvation) day of the month.

In Leviticus 23:34-35, the Lord gives instructions to Moses concerning observing the Feast of Tabernacles. The event was to start on the fifteenth day of the seventh month of the Jewish calendar. They were commanded not to do any servile work. It was a time of rest.

The miracle of the shadow of the sun moving backwards ten degrees in the courtyard of King Hezekiah is one full of numbers. King Hezekiah appeared to be close to the Almighty God. God, through the prophet Isaiah, told him that he was going to die and to set his house in order. God rarely gives a death notice and the date to any man. Hezekiah prayed with tears; faced the wall; and God heard his cry. Isaiah was still within the compound of the ruler of the land. In the courtyard, God told Isaiah to tell King Hezekiah that he would live an additional fifteen years. Within three days, God stated that King Hezekiah would go up unto the house of the Lord. Three is the number of completion. His cured status would be complete for only the healthy were to go to the Temple to worship and the people would see this act. Taking a lump of figs, Isaiah and Hezekiah laid it upon the boil and King Hezekiah recovered. As the Jews require a sign, King Hezekiah was true to that saying. He asked the prophet Isaiah for a sign of his healing. Isaiah gave the king a choice, either let the shadow of the king's sundial go forward ten degrees or let it go backwards ten degrees. It was by nature and earthly laws that the shadow of the sundial go forward, so King Hezekiah requested a supernatural event; that is, that the shadow on the dial go backward ten degrees. It is recorded that Isaiah the prophet cried unto the Lord, and He brought the shadow ten degrees backward (Second Kings 20:11). The number ten symbolizes order. It is an earthly order. Here God overrides the laws of nature and makes the shadow of the sundial go backward by ten degrees. The story of Hezekiah being healed and given fifteen years of life is symbolic of rest. Hezekiah received rest from his deadly disease. The kingdom received a rest from its enemy, the King of Assyria (Second Kings 20:6) and God defended the city. The Israeli soldiers received a rest from the enemy.

In Esther 9:20-22, the Jewish people rested on the fifteenth day of the month from their enemies. Fighting was turned into rest. Sorrow was turned into joy. Mourning was turned into a good day. They sent gifts to one another to show their appreciation to God for giving them rest from Haman and the haters of God's people.

The Sermon

The Rest From God

I. The Master of Rest ~ Matthew 11:28

The Master of rest in the believer's life is the Lord Jesus Christ. The person who is heavy laden with the cares of this world is provided a remedy. The person seeking rest must first come to Christ, and second, learn of Him. Christ promises rest for the soul of the believer.

II. The Method of Rest ~ Mark 6:31

The recommendation of Christ for rest is to come apart from others. The believer in need of rest is encouraged to seek a desert or unoccupied place. Jesus' disciples were so busy that they did not have rest even while eating. There is a time to seek rest from God.

III. The Message of Rest ~ Hebrews 4:9

The writer of Hebrews addresses the message of rest. The children of Israel did not enter this rest because of their unbelief. Unbelief causes worry. The message of rest is one of encouraging the believer to rest in the work of Christ on the cross for them and not to work to be saved. Just as God rested on the seventh day and ceased from His labors, the believer should also rest in God's work of salvation in his life.

CHAPTER 20

THE NUMBER SIXTEEN - "LOVE"

The Source

Sixteen is first mentioned in Genesis 46:18 where it is revealed that Leah bare unto Jacob sixteen souls. In the construction of the fabric walls of the tabernacle, there were sixteen sockets of silver (Exodus 26:25). In Joshua 15:41, there were sixteen cities with their suburban villages. Sixteen villages were mentioned in Joshua 19:22. Johoash the son of Jehoahaz reigned over Israel in Samaria for sixteen years (II Kings 13:10).

The Symbolism

The symbolism of the number sixteen is love. The first mention of the number sixteen in the Scriptures is focused upon the daughter of Laban. Leah had a birth defect. She was tender eyed (Genesis 29:17). This physical condition resulted when the muscles that move the eye are weak, and the person has one eye that pulls to one side or the other. Her appearance because of her eyes was not attractive. Jacob had already fallen in love with Rachel whom he had met near the water well. In bargaining for his wages from Laban, Jacob requested Rachel as his wife and Laban agreed. It is interesting that Jacob set the time of the wages as seven years. Seven symbolizes perfection. The seven years seemed but a few days in the life of Jacob for the love he had for her (Genesis 29:20). When the seven years were accomplished, a feast was given by Laban, and the marriage was set. In the evening of the marriage ceremony and feast, Laban brought to Jacob not Rachel but his firstborn daughter, Leah. The next morning, Jacob discovered the switch that was made and protested unto Laban. Laban explained that according to their customs the first born daughter must be married first (Genesis 29:26). Laban then promised Jacob that if he will work an additional seven years, then he could marry Rachel. The resulting marriage takes place seven years later and then began the rivalry between Leah and Rachel. They vie for the love and affection of Jacob. This entire scenario is observed by God who has love towards Leah and He opens her womb to have children (Genesis 29:31). With the birth of their first son, Leah then thinks that Jacob will love her. Unfortunately, Jacob loved Rachel more than Leah, and this increased the conflict between the two wives. Leah will eventually give birth to sixteen souls (Genesis 46:18) through Jacob. Some were born through her own body and some through her handmaid, Zilpah. The story of Leah shows the love of God for the downtrodden. God states that He made the person who had health problems (Exodus 4:11). Leah had no choice in her birth defect, but God still loved her.

Another time the number sixteen is used distinctly in the Scriptures is in Exodus 26:25. The wooden frame for the walls of the tabernacle were to be joined with silver sockets. The number of these were sixteen. The metal to form these sixteen sockets was silver. Silver is symbolic of redemption which is by love. Gold in the Holy Scriptures is symbolic of faith (First Peter 1:7). The works of the believer will one day be tried by fire. There are six elements mentioned in First Corinthians 3:12, "Gold, silver precious stones, wood, hay, and stubble." These are the attributes of the believer that will be tested by the judgment fire. Silver symbolizes the redemption of the believer which will endure and is found in the sixteen silver sockets of the framework of the tabernacle.

In the great chapter of love in the New Testament, I Corinthians 13, there are sixteen attributes of love. These are:
1. suffereth long;
2. is kind;
3. envieth not;
4. vaunteth not itself;
5. is not puffed up;
6. doth not behave itself unseemly;

7. seeketh not her own;
8. is not easily provoked;
9. thinketh no evil;
10. rejoiceth not in iniquity;
11. rejoiceth in the truth;
12. beareth all things;
13. believeth all things;
14. hopeth all things;
15. endureth all things;
16. never faileth.

Although the use of lists is not entirely recommended in determining the symbolism of a number, there are clear instances in the Bible where the cloaked symbolism occurs. The sixteen attributes of love are the exceptions. God is love (First John 4:16) and is illustrated by this definition of love.

The Sermon

The Three Marks of a Christian

I. Love ~ John 13:35

The first mark of a Christian is to love one another. Christ said this sign would let the world know that the believer was a disciple of His. The commandment to love thy neighbor was complimented by the comments of Christ concerning the love that His disciples would have for one another. You can give without loving, but you can't love without giving.

II. More Love ~ Matthew 5:44

The second mark of a Christian is to love his enemies. It is easy to love those who reciprocate love to a someone, but it takes more love to love those who hate and despitefully use a person. The world needs more warm hearts and less hot heads. More love is one of the true marks of a Christian. To love the world is no big chore. It's that miserable guy next door who's the problem. To love those who hate you is exemplary of the love of God.

III. No Love ~ I John 2:15

The third mark of a Christian is to have no love for the world. The absence of love for the world is a true distinction of a Christian. There cannot be love for God and for the world in the life of a believer. The love for God by a true believer is a fabric which never fades, no matter how often it is washed in the water of adversity and grief. The denial of the things of the world shows the love of the believer for the Heavenly Father.

CHAPTER 21

THE NUMBER SEVENTEEN ~ "VICTORY"

The Source

Genesis 7:11 contains the first mention of the number seventeen when the waters of the deep and the windows of heaven were opened. The story of Noah and his family surviving the world wide flood contains two mentions of the number seventeen (Genesis 7:11, Genesis 8:4). Joseph was seventeen years of age when his trials started (Genesis 37:2). Jacob lived in the land of Egypt under the care of his son, Joseph, for seventeen years (Genesis 47:28). The first Passover was held on the fourteenth day of the month, and three days later on the seventeenth day of the month, the children of Israel crossed the Red Sea and sang a victory song (Exodus 8:27; Exodus 12:6; Exodus 15:1-21). Jeremiah pays the redemption price of seventeen pieces of silver for his family's land and inheritance (Jeremiah 32:6-9; Jeremiah 32:14-15). In Acts 2:8-11, there were seventeen groups of people who heard the wonderful words of God in their own language. In Romans 8:35-39, there are seventeen things listed which are unable to separate the believer from the love of Christ.

The Symbolism

Seventeen is the seventh prime number. The first seven prime numbers are 2, 3, 5, 7, 11, 13 and 17. The number thirteen is symbolic of rebellion and is the sixth prime number. Six is the number of man, and the history of man has been one of rebellion against God. The number seventeen is symbolic of victory and is the seventh prime number. Seven is the number of perfection. When man is not in rebellion against God, but submissive to His will, he is living in victory.

The first mention of the number seventeen is found in Genesis 7:11 where the Bible tells the details of the great flood. God gave the age of Noah when the great flood started and the exact calendar date, the second month and the seventeenth day. At the end of the journey aboard Noah's ark, God again gave exact details of the end of the great flood (Genesis 8:4). The Bible tells us that the ark rested in the seventh month and on the seventeenth day of the month. Clearly by obeying God in building the ark, Noah had victory over the great flood waters of God's judgment on this world. Noah had victory in building the ark, and victory when it landed on Mount Ararat.

At the young age of seventeen, Joseph demonstrates victory in all his trials (Genesis 37:2). Joseph had victory while in the pit, in Potiphar's house, in the prison, and in the palace. Never is there a discouraging word recorded from his lips. He is in control by God's help in every situation he faces. He rises to the throne with Pharaoh, and his father, Jacob, sees his son on the throne in a victorious manner (Genesis 47:28).

The first Passover was conducted on the fourteenth day of the month; and three days later on the seventeenth day of the month, the children of Israel walked on dry ground through the bottom of the Red Sea and sang a victory song (Exodus 15:1-21). It was truly a victory song they sang. Christ was our Passover lamb Who was crucified on the fourteenth day of the month. Three days later on the seventeenth day of the month, Christ arose victorious over death, hell, and the grave (Matthew 28:6).

Jeremiah paid seventeen pieces of silver for his family's land and inheritance which was an act of faith and victory because his payment was at a time when it was evident that the enemy would overtake the promised land. Jeremiah believed that God would allow the children of Israel to return one day to the promised land (Jeremiah 32:6-15; 27). Jeremiah lived with victory at a time when there was no victory in the land of Israel.

On the day of Pentecost (Acts 2:8-11), there were seventeen groups of people from across the Roman empire who had gathered in Jerusalem for the feasts. They heard the wonderful works of God, miraculously, in their own language. They are astonished that these unlearned men could speak to them in this fashion. The greater story is that God allowed the confusion at the tower of Babel (Genesis 11:7) to be dispelled, and they heard the good news as one people with one language. God had wrought a great victory over the obstacle of languages.

In Romans 8:35-39, the believer is told of the seventeen things which are unable to separate the believer from the love of God. Victory in a believer's life and the love of Christ are closely associated. The believer of Jesus Christ has victory in this life. The idea associated with the number seventeen in Scripture is victory.

The Sermon

The Victory of the Believer

I. The Victory Through Salvation ~ I Corinthians 15:54

At the moment of salvation, the new born believer has won the victory through Christ over self, sin, Satan, and death. Death is swallowed up in victory the moment the repentant person turns his life to Jesus Christ.

II. The Victory Through Sanctification ~ Romans 8:2

The believer who conforms his life to the word of God and the Holy Spirit obtains a victory over the world. The believer is to be in submission to the will of God (Romans 8:2).

III. The Victory Through Surrender ~ Romans 8:5

The believer in the everyday walk finds victory by surrendering to the will of God. The will of man is strong and often denies the believer a victorious walk, but a total surrender to God provides a victorious life (Romans 8:5).

CHAPTER 22
THE NUMBER EIGHTEEN ~ "BONDAGE"

The Source

Eighteen is first mentioned in the Bible in Judges 3:14 where the children of Israel served in bondage to Eglon, the king of Moab. Eighteen is also mentioned in Judges 10:6-9. King Josiah found the book of God in Second Kings 22:1-20 and was informed of the history of the people of Israel for the past eighteen years. Nebuchadnezzar, in his eighteenth year, carried Israel into captivity. There were eighteen souls that perished at the tower of Siloam (Luke 13:1-5). In Luke 13:11-17 the story is told of the woman who had a spirit of infirmity for eighteen years.

The Symbolism

The symbolism of the number eighteen is bondage. Bondage is the state of being greatly constrained by circumstances or obligations. Bondage is the state of being a slave whether to another person or to some influence. The first mention of the number eighteen in the Bible (Judges 3:14) gives an excellent idea of the symbolism of the number. In this verse, the children of Israel served Eglon the king of Moab eighteen years. Again in Judges 10:6-9, the children of Israel are in bondage to the Philistines and the Ammonites for eighteen years.

On the eighteenth year (Second Kings 22:3) of his reign, King Josiah decided to repair the House of God. Then Hilkiah, the high priest, (Second King 22:8) found the book of the law in the house of the Lord. Through the reading of the book of the law, King Josiah realized that the children of Israel had been in bondage to idolatry for the eighteen years of his reign. In Jeremiah 52:29, Nebuchadnezzar, in his eighteenth year, carried Israel into slavery or bondage.

Jesus stirs the imagination and thoughts of his followers in Luke 13:1-5 by presenting a question. They had related to Him the death of some Galileans whose blood Pilate had mingled with their sacrifices. The followers had insinuated that these were sinners because of the means of their deaths. Christ relates that these were not sinners, but that except his followers repent, they shall all likewise perish. Jesus reminds his followers of the structural accident in Siloam when the tower of that town fell and killed eighteen residents. Jesus asks, "Think ye that they were sinners above all men that dwelt in Jerusalem?" Jesus agains reminds them that except ye repent, ye shall all likewise perish. In John 8:34, Jesus stated that whosoever commits sin is the servant or in bondage of sin. Sin holds people as chains hold the slave.

Luke immediately illustrates the principle of bondage by the telling the story of the woman who had an infirmity for eighteen years. Before healing the woman of her ailment, Jesus said, "Woman, thou art loosed from thine infirmity." The problem that arose centered around this healing occuring on the Sabbath. The ruler of the synagogue arose with anger and denounced the healing on the Sabbath. Jesus boldly called him a hypocrite because the ruler of the synagogue would have never made such a statement if one of the members of the synagogue had loosed his ox or his ass from the stall and led it away to water it. Jesus boldly stated that this woman had been under bondage for eighteen years (Luke 13:15-16). The greater story was the bondage of the children of Israel to the Mosaic law, especially the Sabbath day.

The Sermon

Bondage to Sin

I. The Sinner Is The Servant of Sin ~ Galatians 4:3

The person who sins becomes the servant of sin whether or not the person is saved. The chains of sin are not prejudicial. They will attach themselves to any person (Galatians 4:3). The sinner is the servant of sin.

II. Believers Can Be The Servant of Sin ~ Galatians 4:9

Those who knew God and turned back to the beggarly elements of the world are declared to be in bondage again (Galatians 4:9). Believers can be servants of sin whenever they turn to the world for their pleasure and desires.

III. Bondage to Sin Is Liken Unto the Yoke of Oxen ~ Galatians 5:1

The apostle Paul urges the Galatians to stand fast in the liberty that Christ has given. He urges them to watch and not be entangled again with the yoke of bondage (Galatians 5:1). Peter warns that whomever or whatever you serve, you are brought into bondage to that person or thing. Paul considered himself to be the servant of God (Romans 1:1).

CHAPTER 23
THE NUMBER NINETEEN ~ "WISDOM"

The Source

The number nineteen is found only two times in the Bible. The first mention is in Joshua 19:38, and the second reference is in II Samuel 2:30. In the first reference, Joshua tells of the nineteen fenced cites in the inheritance of Naphtali. In the second reference, nineteen of David's warriors and Asahel died in a battle whereas the enemy lost three hundred sixty men. Nineteen is the eighth prime number. The number eight in the Bible symbolizes a new beginning.

The Symbolism

The symbolism of the number nineteen is wisdom. Wisdom is a difficult thing to find in the lives of people. In the two verses where the number nineteen is found, there are two conflicts. In the first verse is the defensive wisdom a person needs, that is, building a fence around their cities (Joshua 19:38). In the second verse is the offensive wisdom a person needs, that is, having fewer losses than the enemy (II Samuel 2:30). The number nineteen is the eighth prime number and symbolizes that wisdom is a new beginning in the life of a believer. When a believer acquires wisdom from God for daily living, whether defensive or offensive, it is a new beginning.

Joshua 19:32-39 tells of the inheritance of the children of Naphtali, one of the children of Jacob. Naphtali was described by Jacob as "a hind let loose: he giveth goodly word" (Genesis 49:21). Good words are a sign of wisdom. Proverbs 18:4 states, "The words of a man's mouth are as deep waters, and the wellspring of wisdom as a flowing brook." Joshua 19:35 tells of the nineteen fenced cities. There is great wisdom in fencing a city, that is building a walled city. A wise person puts a fence about his life, his family, his church, and his faith in God. This is defensive wisdom.

On the offensive side of wisdom, a wise person wisely chooses his battles. In II Samuel 2:30, only nineteen of David's servants and Asahel died in the battle, whereas three hundred sixty of the enemy died. Wisdom prevented the loss of life. The men of David were wiser than the men of the enemy and wisely fought their battles. A wise person wisely chooses his battles.

Although the number nineteen is not found in the New Testament, Joshua 19:32-38 points the reader to the Sermon on the Mount. The inheritance of Naphtali described by Joshua contained Capernaum and Bethsaida on the north side of the sea of Tiberias. The area described by Joshua as the inheritance of Naphtali was the locality in which Jesus did many mighty works, and this was the place where Christ delivered His Sermon on the Mount (Matthew 5-7). In this sermon, Jesus Christ (Matthew 7:24-29) ended his discourse with the reference to the wise man and the foolish man. The wise man is the one who hears the words of Jesus and builds their life upon them. The greatest statements of wisdom were delivered by Jesus in the land region inherited by the descendants of Naphtali. The idea associated with the number nineteen in the Bible is wisdom. Though there are only two references in the Bible, we have examples of defensive and offensive wisdom. Nineteen is symbolic of wisdom.

The Sermon

Wisdom

I. What Is Wisdom? ~ I Kings 3:16-28

Wisdom is knowledge in action. Wisdom is the ability to know the difference in a situation. King Solomon was given an unusual gift, wisdom. The classic story in First Kings 3:16-28 unveils the wisdom of King Solomon. The story reveals that Solomon is asked to determine the true mother of the remaining child. After instructing one of the guards to divide the child with the sword, the false mother agrees to the bisecting of the child, whereas the true mother pleads that Solomon not kill the child by dividing it. Wisdom was demonstrated by Solomon by his ability to know the difference between the false mother and the real mother. After speaking his judgment of dividing the child with the sword, Solomon kept silent and observed the reactions of the two mothers. Their reactions allowed him to determine the true identity of the mother. Knowledge is knowing a fact. Wisdom is knowing what to do with that fact.

Wise men think without talking, but fools reverse the order. Wisdom is knowing what to say and the correct time to say it. Studying the Bible, especially observing how Jesus conducted Himself, and putting those principles in place in a person's life will lead to a wise path. One can buy education, but wisdom is a gift from God.

II. Who Is a Wise Person? ~ Deuteronomy 4:5-6

The Old Testament reveals that a wise person is one who has been taught and kept the statutes and judgments of God (Deuteronomy 4:5-6). The New Testament shows that a wise person is one who hears the teachings of Jesus and keeps them (Matthew 7:24-29). A person should have enough education so he doesn't have to look up to anyone, but he should also have enough wisdom not to look down on anyone. Some folks are wise and some are otherwise.

III. Wisdom Comes From God. ~ James 1:5

The first sign of wisdom is to be saved and escape hell (Psalm 111:10). Having a holy reverence for God is the beginning of wisdom. God is the One who gives wisdom (Proverbs 2:6). James tells us that if a person lacks wisdom, he is to ask God (James 1:5). Jacob described Naphtali as having goodly words. These were wise words. The courage to speak must be matched by the wisdom to listen. Knowledge is knowing a fact, while wisdom is knowing what to do with that fact. Knowledge comes by taking things apart, but wisdom comes by putting things together. Wisdom comes from God.

CHAPTER 24

THE TWIN PRIME NUMBERS

The Prime Numbers That Are Twins

The prime numbers are the numbers which have only two factors, one and itself. A factor is any number which will divide evenly into another number. For instance, two and three are factors of the number six, because they will divide into six evenly with nothing remaining. Twin primes are two prime numbers located near each other and separated by one number. The exception is the first two prime numbers, two and three. The next twin prime numbers are 3 and 5; 5 and 7; 11 and 13; 17 and 19. Mathematicians have proved that there is an endless quantity of twin prime numbers. One of the great unsolved problems in mathematics is to find a mathematical formula to predict the twin prime numbers.

The Interdependence of the Symbolism of Twin Prime Numbers

A study of the prime numbers and their Biblical symbolism reveal an interdependence upon each other. The first twin prime numbers are two and three. Two symbolizes witness and three symbolizes complete. The person who accepts the witness of the Word of God becomes complete in Christ. The next twin prime numbers are three and five. Three symbolizes complete and five symbolizes grace. When a person is complete, the grace of God is fully realized in their life. Five and seven are twin primes. Five symbolizes grace and seven symbolizes perfection. Only through the grace of God can man be perfect. Eleven and thirteen are twin prime numbers. Eleven symbolizes disorder or judgment. Thirteen symbolizes rebellion. Disorder and the need for judgment is linked to rebellion in a person's life.

The twin prime relationship between seventeen and nineteen is undeniable. Seventeen is symbolic of victory. Nineteen is symbolic of wisdom. The believer in Christ should possess these two attributes to finish their Christian course in an honorable manner. Noah possessed the victory (17) by surviving the great flood, but he did not exhibit wisdom (19) by drinking fermented wine. It caused a divide in his family and a loss of victory. David was blessed with victory (17) as the King of Israel but he did not use wisdom (19) in his relationship with Bathsheba. It almost cost him the kingdom much less the death of an honorable man. Demas is praised by the Apostle Paul in Scripture (Colossians 4:14). It is clear from these that he had victory (17) in the Christian walk. It is a sad report though when Paul reports that Demas (II Timothy 4:10) had forsaken him. Demas did not possess the wisdom (19) to keep himself separate from the world. Victory and wisdom do walk together because they are agreed (Amos 3:3). There is a relationship between the twin prime numbers and their Biblical symbolism. There is a divine design in the digits. God has created the number system and has placed in the Scriptures the numbers in the right location to reveal their symbolism. The number nineteen symbolizes wisdom and wisdom is difficult to find in a person's life. Proverbs 25:2 says, "It is the glory of God to conceal a thing, but the honour of kings to search out a matter."

CHAPTER 25
THE NUMBER TWENTY ~ "REDEMPTION"

The Source

The first mention of the number twenty in the Bible is in Genesis 18:31 where Abraham proposed that God spare Sodom if there were twenty righteous people there. In Genesis 31:38, Jacob presents his proposal to Laban for his years of labor. Instead of death, Joseph is sold to the Midianites for twenty pieces of silver (Genesis 37:28). The number twenty is found in Exodus 30:11-15 where the price for the offering to the Lord is noted. The lower age limit to give a mandatory offering to the Lord was twenty (Exodus 30:14). The age of mandatory military service was set at twenty (Numbers 1:24). The ark of the covenant was possessed by the Philistines for twenty years (First Samuel 7:2). It took twenty years for King Solomon to complete the construction of two houses - the house of the Lord, and the king's house (First Kings 9:10). The number twenty is found one time in the New Testament in Acts 27:28.

The Symbolism

In negotiating with God concerning the number of righteous persons living in Sodom that would gain God's grace in sparing Sodom from destruction (Genesis18:31), the number twenty is found as the next to the last order of the amount inquired. This is the first mention of twenty in the Bible. In Genesis 31:38, Jacob labored twenty years to redeem his wives, Leah and Rachel, and the cattle in his possession. For the atonement of their souls, each male, age twenty or more, had to give twenty gerahs in silver for their redemption.

Joseph is sold to the Midianites for twenty pieces of silver (Genesis 37:28). Silver is a type of redemption. The idea associated with the number twenty in this passage and other verses in the Holy Scriptures is redemption.

In Acts 27:28, the sailors sounded, or determined the depth of the water, by dropping a weighted rope, and at their first sounding found the bottom to be twenty fathoms. The next time the sounding was held, the bottom was found to be fifteen fathoms. Fifteen symbolizes rest. The reason for the sounding of a water-born vehicle is to save it from grounding and thus its destruction. The number twenty symbolizes redemption. Their efforts in determining the bottom of the ocean was to save the vessel from destruction.

The Sermon

Redemption

I. The Clarification of Redemption ~ Leviticus 25:24

Biblical redemption is the action of saving or being saved from sin. Adam had "sold" our possession into the hands of Satan, and like the year of Jubilee, then sinners can be redeemed by God if they will turn to Him for salvation. During the year of Jubilee, land and houses were able to be bought by the original owner from the person who had obtained the estate. Redemption is the return to the original owner of a commodity. The human race was originally in the hands of God and can be redeemed by Him if the sinner is willing to repent and be saved. Redemption is not buying something that never was owned, but rather buying something that was once possessed by the purchaser.

II. The Call for Redemption ~ Romans 10:13

Man fell in sin, and because of this act of disobedience, there is the need for redemption. The first act in redemption is for the person needing this action to admit his need (Romans 3:23). The second step is to believe on the Lord Jesus Christ (Acts 16:31). The third step is to call upon God to be saved from his sin (Romans 10:13). God calls upon everyone to turn back to Him through the preaching of the Gospel of Jesus Christ.

III. The Cost of Redemption ~ I Peter 1:18-19

Saint Peter gives a clear analysis of the cost of redemption of the soul. It is not with corruptible things (First Peter 1:18 - 19), but with the precious blood of Christ as of a lamb without blemish and without spot. Only through the blood of Christ can the sin of a person be forgiven and the person experience redemption. It is Christ who paid the price to buy back the title deed of creation which was sold by Adam in the garden of Eden. The price has been paid for every person. God waits upon man to repent and to turn to Him for salvation.

CHAPTER 26
THE NUMBER TWENTY-ONE ~ "EVIL"

The Source

The first mention of the number twenty-one is in Second Kings 24:18 where the life of King Zedekiah is described. Details of his life are also described in II Chronicles 36:11-12 and in Jeremiah 52:1-11. In Daniel 10:13, the prince of the kingdom of Persia withstood the angel of God for twenty-one days in bringing an answer to Daniel's prayers. Although the number twenty-one is not explicitly found in the New Testament, there is a location which contains twenty-one references to sin (Second Timothy 3:2-5).

The Symbolism

The association of the number twenty-one with the ideas found in the Bible conclude that the symbolism of this number is evil. In the conquest of the land of Israel, Nebuchadnezzar carried away King Jehoiachin to Babylon (Second Kings 24:15). In his place, Nebuchadnezzar installed the uncle of King Jehoiachin whose name was Mattaniah. Nebuchadnezzar changed his name to Zedekiah (Second Kings 24:17). Zedekiah was twenty-one when he assumed the position as ruler and reigned for eleven years. Eleven is the number of judgment. The Bible states the position of Zedekiah in simple terms,

"And he did that which was evil in the sight of the Lord." Even though Zedekiah was appointed by Nebuchadnezzar, he rebelled against him in the ninth year of his reign (Second Kings 25:1), Nebuchadnezzar moved to replace King Zedekiah from his throne. For two years the war raged, and the army of Nebuchadnezzar starved the city of Jerusalem into subjection (Second Kings 25:3). King Zedekiah was captured while fleeing the city of Jerusalem on the road to Jericho. In the eleventh year of his reign, the last thing King Zedekiah observed was his sons being killed, then the Chaldeans put out his eyes. Eleven symbolizes judgement and twenty-one symbolizes evil.

Daniel fasted and prayed for twenty-one days in seeking answers from God concerning the future (Daniel 10:13). The evil that is associated with the number twenty-one in this passage is that the prayer of Daniel was hindered by the wicked underworld. Satan hates the Word of God and will stop at nothing to hinder its progress.

The Apostle Paul instructs Timothy (Second Timothy 3:2-5) on the decadent behavior of men in the last days. There are twenty-one distinct characteristics given which are evil. Second Timothy 3:2-5 states, "In the last day perilous times shall come, for men shall be:

1. lovers of their own selves,
2. covetous,
3. boasters,
4. proud,
5. blasphemers,
6. disobedient to parents,
7. unthankful,
8. unholy,
9. without natural affection,
10. truce breakers,
11. false accusers,
12. incontinent,
13. fierce,
14. despisers of those who are good,
15. traitors,
16. heady,
17. highminded,
18. lovers of pleasure,
19. more than lovers of God,
20. having a form of godliness,
21. but denying the power thereof: from such turn away." Any one of these is evil within itself, and the totality of this behavior is very evil.

The Sermon

The Results of an Evil Life

I. The LSD of Sinners ~ James 1:15

Lust brings forth sin, and sin brings forth death (James 1:15). Lust is from an evil source. Sin is the result of lust which becomes evil in the life of the person. The eventual end of sin is death. The LSD of sinners is lust, sin, and then death. LSD is a drug which distorts the user's surroundings. Lust, sin, and death are the roadsigns along an evil path of the sinner.

II. Sin Leads to Evilness ~ James 1:15

Sin in the life of a person may not be evil at the onset, but it is on the path to being evil. Sin at first may be simply pleasure in sin which will last for a season. If left unchecked, the sin in a person's life could become evil. A serial killer first started the path to being a horrible murderer by allowing sin to remain.

III. Evilness Leads to Death ~ Romans 6:23

Sin allowed to remain in a person's life further corrupts the person to become evil. Few men have the character to resist the further steps of sin which is evil. The life of Zedekiah (II Kings 25:1) is a sad example of where evilness led to death. Zedekiah had a great kingdom and could have lived in peace with Nebuchadnezzar, but his evilness in rebellion led to seeing his sons killed before his eyes, and then the loss of vision.

CHAPTER 27

THE NUMBER TWENTY-TWO - "LIGHT"

The Source

The first mention of the number twenty-two in the Bible is in Joshua 19:30-31 where the twenty-two cities in the inheritance of Asher are mentioned. Jair, a Gileadite, judged Israel twenty-two years (Judges 10:3). Jeroboam reigned twenty-two years (First Kings 14:20). Zadok was a young man of valor who served along with twenty-two captains of his father's house (I Chronicles 12:28).

The Symbolism

The passages in Joshua 19:30-31 describe the twenty-two cities which are encompassed in the inheritance of Asher. The name Asher means level, right, happy, to go forward, to be honest, to be proper, and to be blessed. This is a person who lets their light shine like sunshine. The twenty-two cities in the inheritance of Asher were points of light in the promised land. In the New Testament, light and cities are mentioned in the beginning and end of the books (Matthew 5:14; Revelation 21:23). The first time *light* is used in the New Testament is in Matthew 4:15-16 where the region described "by the way of the sea" is the region of the twenty-two cities denoted in Joshua 19:30.

The Hebrew language contains twenty-two letters. The majority of the Word of God in the Old Testament consists of these twenty-two letters and has brought light to the world of the ways of God. Psalm 119 is divided into twenty-two sections, with each section starting in the Hebrew language with a corresponding letter of the Hebrew alphabet. Psalm 119:105 states, "Thy word is a lamp unto my feet, and a light unto my path." With twenty-two symbols, God has given us the majority of the Old Testament which is a great light for man.

A candlestick was placed in the holy place of the Tabernacle (Exodus 25:31-34). It had three branches on each side which consisted of six branches. Each branch had three bowls which equaled eighteen bowls. Along with the eighteen bowls, the candlestick shaft had four bowls, thus the candlestick had a total of twenty-two bowls in which olive oil was poured to provide fuel for the flame from the six branches and the center branch. The olive oil is symbolic of the Holy Spirit and the twenty-two bowls are symbolic of light.

The Sermon

The Light of the World

I. Christ Is the Light of the World - John 8:12

Jesus stated that He is the light of the world. The person who follows Christ will not walk in darkness and sinfulness of the world. Christ is like the sun and the believer is like the moon. One produces the light and the other reflects the light. The believer receives his light from Christ. Without Christ in his life, the believer would be as ineffective as the moon without the light of the sun. Christ, not another man, is the true light of the world.

II. The Believer is the Light of the World - Matthew 5:14

Christ proclaims that the believer is like a city set upon a hill. It cannot be hid. The purpose of the light is to dispel the darkness. Christ stated that men do not light a candle and place it under a basket, rather they place it in a prominent place in the house to drive away the darkness. The light helps those who live in the house. The believer helps those who live in the world by giving them the Gospel.

III. Put on the Armor of Light - Romans 13:12

The Apostle Paul encouraged the Roman believers to put on the armor of light. The reason for this urgency was that time was passing. The night is far spent, and soon it will be the sunrise. The believer must get up and be about the Father's business. Paul tells them that they have been given the light of the glorious Gospel. The works of darkness survive because the believers do not let their lights shine and repel those evil works.

CHAPTER 28

THE NUMBER TWENTY-THREE ~ "DEATH"

The Source

The number twenty-three has a peculiar place in the line of numbers. It is a prime number. It is one of the special numbers that nothing will divide into it equally except one and itself. The prime numbers in the number system are the building blocks of it. It is a foundational basic number. The number twenty-three is found first in Judges 10:2.

Tola is a man who was a judge (Judges 10:2). He judged for twenty-three years. He, in essence, was the head supreme court judge. He at some times delegated judgments. He had a circuit and traveled to different parts of Israel. When he arrived at a city such as Jerusalem, people would come with their grievances to him. He would listen all day, and cast judgment on what had to be done. Then he would travel to another city and do the same thing. He was a circuit judge for twenty-three years. The name Tola means "worm or maggot." This was the crimson grub used to dye cloth crimson or scarlet color. He is a grub. His name meant that he was a grub or a maggot. Job said that even if skin worms would devour his flesh, yet will he seek God. Psalms 22:6 is a reference to the worm.

The Symbolism

What does grub or maggot bring to our remembrance? It reminds one of death. The number twenty-three is symbolic of death. The Scripture even tells us where he was buried. He was buried in Shamir. The word Shamir means a pricking, a thorn, as a gem or diamond scratches a stone, and here is a hidden reference to a grave marker. His name means "worm", and the place where he is buried means "grave marker". It tells us in twenty-three years he died. The number twenty-three is a prime number, a foundational basic number. It is appointed unto man once to die, and then the judgment, as Hebrews 9:27 states. The first mention of the number twenty-three in the Bible does not have a good connotation. There is death all around. He is a maggot. He dies. He is buried in a place called the grave marker. Death is found throughout this verse in Judges 10:2.

In II Kings 23:31-32, it states that Jehoahaz was twenty and three years old when he became king and did evil. Evilness is a sign of death. The wages of sin is death (Romans 6:23), and when someone does evil, it is an invitation to death and corruption into a person's life. Evilness brings forth death, as it did to this king who was twenty-one when he became ruler.

In Esther a decree was issued to put the Jews to death. It was their Persian holocaust. On the twenty-third day of the month, King Ahasuerus signed the decree stating he cannot reverse the other decree, but he could sign one allowing the Jews liberty to take up arms and defend themselves. Esther 8:8-11 clearly revolves around the theme of death.

The Psalms were written in order and had a number attached to them from the writing of the original in Hebrew. The middle of the twenty-third Psalms states, "Yea, though I walk through the valley of the shadow of death." In a great and inspiring Psalms like the Twenty-third Psalms is found death.

The Sermon

The Specter of Death

I. Death Is Waiting ~ Romans 1:32

Although the number twenty-three is not explicitly found in the New Testament as the name of God is not found in the book of Esther, there is the number twenty-three in the shadows of New Testament teachings. Romans 1:28-31 lists twenty-three characteristics of people that God had turned over to a reprobate mind. The list includes doing those things which are not convenient; being filled with all unrighteousness, fornication, wickedness, covetousness, maliciousness; being full of envy, murder, debate, deceit, malignity; whisperers, backbiters, haters of God, despiteful, proud, boasters, inventors of evil things, disobedient to parents, without understanding, covenant breakers, without natural affection, implacable, and unmerciful. Romans 1:32 states that those who, knowing the judgment of God, commit such things are worthy of death, not only do the same, but have pleasure in them that do them. Although the number twenty-three is not found directly in the New Testament, there are twenty-three things that bring forth death. The number twenty-three reveals this death.

Romans 6:23 tells us that the wages of sin is death. People who commit those things and never repent, die a continual death and never receive the gift of God which is eternal life. Jesus talked about the second death. The first death will be in this life, and the second death will be living in a continual, burning hell, forever experiencing death. In the lake of fire, death and hell will be poured. It is a terrible thing to be in a continual state of death. The eternal torment in the lake of fire will not only be the torment by fire, but the realization of the nonexistence of hope of escaping. All hope is gone in the lake of fire. Death is not a period but a comma in the story of life.

II. Death Reigns - Romans 6:23

Romans 7:9 states, "For I was alive without the law (number 10 which is order) once: but when the commandment came, sin (number 13 which is rebellion) revived, and I died (number 23 which is death)." The number ten, order or the law, plus the number thirteen, which is rebellion or sin, equals the number twenty-three, which is death.

The best way is to locate the symbolism of a number in Scriptures is to study where it is mentioned in the Bible. In Judges 10:2, a man called Tola, ruled for twenty-three years, died and was buried. Death is when the body quits functioning. The facts of death are simple. It is appointed unto man once to die, and then the judgment (Hebrew 9:27). In Ecclesiastes it states there is a time to live and a time to die. For every person, there is a time span which includes our birth, life and death. Nobody knows when they are going to die, and they have to be told when they were born.

Not only is there physical death, but there is spiritual death. Spiritual death is found in the book of Genesis where God warned his creation, Adam and Eve, not to eat the fruit. The day they do, they shall die. They did not physically die that day, but they spiritually died that day. God rushed angels in around the tree of life. He surrounded it with flaming swords so Adam and Eve could not eat the fruit of the tree of life and be in an eternal state where they never could escape death. God had to make a way for them to physically die.

When Christ said you must be born again, he was talking about the spiritual man. Nicodemus did not understand and thought it not possible to be born again. Christ was not talking about the physical birth, he was talking about the spiritual birth.

There is a physical death and a spiritual death. The person who has been born again will not die again, because that person has been born of incorruptible seed which is the Word of God.

III. Death Has Been Defeated ~ I Corinthians 15:55 - 57

The number twenty-three is not the number of the believer. Christ died on a cross and overcame death. He had power over death. The believer does not need to be afraid of death because Christ died. He willingly laid down his life. He gave up the ghost. He would never have died if He had not willed Himself to die. He willingly died and conquered death. They put Him in the tomb for three days, and the grave could not hold him. The three pillars of Christianity are the incarnation (the Son of God born of a virgin), the crucifixion (He died on a cross for our sins), and the resurrection (He came out of the grave). Because He arose, number twenty-three has no effect on the believer of Jesus Christ

CHAPTER 29

THE NUMBER TWENTY-FOUR ~ "PRIESTHOOD

The Source

The first mention of the number twenty-four in the Bible is in Numbers 7:88 where it mentions that twenty and four bullocks were sacrificed for the dedication of the altar after it had been anointed. In Numbers 8:19, there are two duties of the priesthood. The priests were to do the service of the children of Israel in the Tabernacle. Those men entered the holy area, received from the person or family the sacrifice, and offered it to the Lord. No one else was to go into the Tabernacle or Temple. Everybody else was excluded. If a person was not of the lineage of Aaron that person could not come into the holy place. The priests were intercessors between the people and a holy God by bringing the people's sacrifice unto God. In Numbers 8:19 it says to make an atonement for the children of Israel. If a person had sin, that person could not bring their turtle dove, their lamb, or their offering into the Tabernacle or Temple. They had to bring it to the priest, tell him their sin or problem, and the priest would take the sacrificial offering. The priest was an intercessor between the people and God. The priest would take the sacrifice and burn it on an altar. After their confession the priest brought it in and confessed what had been wrong. This offering would be an atonement for the children of Israel. Once a year the Passover lamb was brought for the entire nation of Israel. It was a blanket coverage for sin of the nation of Israel. The priesthood was very important in the work of God.

In I Chronicles 24, the priesthood was established with the sons of Aaron as priests. Abihu and Nadab, two sons of Aaron, died because they brought in strange fire. God had lit the fire of the altar from heaven, and that fire was kept alive. God is not going to accept a sacrifice offering in any manner except which He had authorized. Aaron's sons, Eleazar and Ithamar, executed the priest's office after the death of Abihu and Nadab. Eleazar had sixteen sons, and Ithamar had eight sons, which is twenty-four. Twenty-four men are listed who have their family in charge of the priesthood. They served for two weeks each year and were divided into twenty-four courses.

The father of John the Baptist, Zacharias, was in the course of Abia (Luke 1:5), which was the eighth course, or new beginning. John the Baptist came out of the eighth course, and he was definitely a new beginning. John the Baptist could have been the high priest. John the Baptist was the true priest of God, and truly he was dedicated to God. He did not stay in the priesthood seminary. He was a descendent of Aaron. When

Mary, the mother of Jesus, visited Elizabeth, John the Baptist jumped in her womb at the sound of Mary's voice. Mary and Elizabeth were cousins. Mary was not of the course or lineage of Aaron, and neither was her espoused husband Joseph. They were of the tribe of Judah. They were not of the priesthood from Aaron, but Elizabeth was.

Other Scriptures where twenty-four is used is in Daniel 10:4 where Daniel has a vision. Also Haggia 1 and Zechariah 1 deal with the priesthood, the Temple, and theTabernacle.

In the book of Revelations, the number twenty-four is found. It is not mentioned anywhere else in the New Testament except in the book of Revelations. In Revelations 4, it states there were four and twenty seats, and upon the seats were four and twenty elders sitting, clothed in white raiment, and they had on their heads crowns of gold. They are before the throne praising God. In verse ten, four and twenty elders fall down before Him that sat on the throne and worship him that liveth forever and ever, and cast their crowns upon the throne saying, "Thou art worthy, oh Lord, to receive glory and honor and power, for thou hast created all things and for thy pleasure they are and were created." In Revelations 5:8, it states that when He had taken the book, the seven sealed book, the four beasts and the four and twenty elders fell down before the Lamb, having every one of them harps, golden vials full of odors which are the prayers of saints. They sang a new song saying, "Thou art worthy, take the book, to open the seals thereof, for thou wast slain and has redeemed us to God by thy blood out of every kindred and tongue and people and nation".

In Revelations 11:16, the four and twenty elders, which sat before God on their seats, fell upon their faces, and worshipped God, saying, "We give thee thanks, O LORD God Almighty, which art, and wast, and art to come; because thou hast taken to thee thy great power, and hast reigned." Christ will take control of all the kingdoms of the world, and He is going to reign. In Revelations 19:4, the four and twenty elders and the four beasts fell down and worshipped God who sat on the throne. The prophetic book of Revelation appropriately contains the number twenty-four for it is the book of the divine priesthood in the New Testament. Peter declared that every believer was in a royal priesthood (I Peter 2:9).

The Symbolism

The idea associated with the number twenty-four in the Scriptures is the priesthood. The surviving sons of Aaron had twenty-four sons who became the servants of God at the Tabernacle and the Temple. The twenty-four elders in Heaven around the throne of God were clearly worshipping God as they proclaimed His greatness and holiness.

The Sermon

The Priesthood of the Believer

I. The Christian Believer and the Priesthood ~ Revelation 20:6

Revelations 20:6 states that the believer is the priesthood. The saints of God will be the priests of God and Christ, and shall reign with him a thousand years. Every saved person will have the job and duty as the twenty-four elders did of worshipping God and praising Him.

In I Peter 2:9, Peter said unto the people of God, "That they are a chosen generation, a royal priesthood, a holy nation, a peculiar people, and should show forth the praises of him who have called them out of darkness into His marvelous light". The believer has been called out of darkness into light, and he is a chosen person. Believers of Christ are a royal priesthood. The middle wall of petition is broken down, and the believer can enter into the presence of God personally before the throne of God. The believer has the power and ability to meet God in prayer and ask him for the things needed. The believer is told to approach the throne of God with boldness and he will find help in the time of need. The believer of Christ is a royal priest.

II. The Royal Priesthood ~ I Peter 2:9

Peter declared the believers as a royal priesthood. There was a twenty-four priesthood schedule, and twenty-four bullocks were sacrificed pointing to the priesthood. Twenty-four men were chosen and given two weeks at a time to serve God. The believer of Jesus Christ is a royal priest of God.

III. The Work of the Priest ~ Leviticus 10:3

God declared that He would be sanctified through the men that came nigh to Him and offered the sacrifices. Men of the tribe of Aaron were called into the priesthood at the age of twenty-five. They were required to shave their head, their beard, and their body. They served until they were fifty years of age. When they became fifty years of age, they could no longer be in the course to serve the Tabernacle or the Temple. They did not retire. They stepped aside and helped the younger men in the priesthood. It was sometimes hard work being a priest. Someone had to cut the wood. They drew lots, and sometimes the lot fell that the priest would cut the wood, because wood was needed for the altar. Someone had to slaughter the animals. There was work in being a priest of God.

CHAPTER 30

THE NUMBER TWENTY-FIVE ~ "FORGIVENESS"

The Source.

In Numbers 8:24 is the first mention of the number twenty-five where the Scripture speaks of the age of the priests of the children of Levi. The men's eligibility required that they had to pass a physical test. They could have no blemishes in their bodies, and had to shave their faces and heads. There were different duties to serve in the temple, but they had to be at least twenty-five years of age. When they reached fifty years of age, they could not serve in the Tabernacle, but they served in the Tabernacle of the congregation.

Evilmerodach showed forgiveness to King Jehoiachin on the twenty-fifth day of the month (Jeremiah 52:31). Evilmerodach brought King Jehoiachin out of the prison. He spake kindly unto him. He gave him a throne above the other kings in his rule. Ezekiel described the measuring of the Temple and the number twenty-five occurred often (Ezekiel 40:21-36). Nehemiah noted the day that the wall around Jerusalem was completed was on the twenty-fifth day of the month. (Nehemiah 6:15-16).

The number twenty-five is found one time in the New Testament in John 6:19 where Jesus walks on water. Great peace was found in this Scripture by the disciples when they knew in this troublesome time of their life that Christ was still in command.

The Symbolism

The number twenty-five is symbolic of forgiveness. The number twenty-four is symbolic of the priesthood. After the priesthood symbolized by number twenty-four, then comes forgiveness. It is interesting that the number twenty-five, forgiveness, is linked with the number twenty-four, the priesthood. A person cannot experience forgiveness until there has been a change in their life, or an intercessor.

The number twenty-five symbolizes the forgiveness of sin. Five times five equals twenty-five. If grace, which is symbolized by the number five, is multiplied by grace again, then twenty-five, forgiveness, is obtained. Christ cried out from the cross, "Father, forgive them for they know not what they do." In Jeremiah 52:31, Evilmerodach, king of Babylon, forgave Jehoiachin, king of Judah. Evilmerodach brought Jehoiachin out of prison, restored his clothing, and permitted him to eat at his table. This event occurred on the twenty-fifth of the month.

The Sermon

I. Forgiveness is a Work of God - Jeremiah 52:31

Ezekiel describes the measuring of the Temple and the number twenty-five is used numerous times. Nehemiah 6:15 says the wall around Jerusalem was finished on the twenty and fifth day of the month Elul, in fifty and two days, and it came to pass that when all the enemies heard and saw these things, they were much cast down in their own eyes. When the enemies of Nehemiah saw the wall had been erected, they had to come to the conclusion that this work was wrought of God. The wall was finished on the twenty-fifth day of the month. Forgiveness in a person's life is perceived by other people as a work of God. For God to change somebody and forgive them of their sins, it clearly has to be an act of God. If somebody gets forgiveness from God, it will change their life like it did King Jehoiachin. King Jehoiachin was forgiven by King Evilmerodach who released him from prison and gave him a reprieve. No one can reform themselves to forgive the way that God forgives. A Christian has two marks: giving and forgiving.

II. Forgiveness Brings Peace - John 6:19

The number twenty-five is found one time in the New Testament. John 6 deals with the account of Jesus walking on the water to their ship. In the book of John, there are around ten miracles mentioned, and the miracles are put in order. The fifth miracle in the book of John is Jesus is walking on the water, and He brings peace to the disciples in the boat. This miracle deals with peace.This is one of the attributes of salvation. One of the characteristics of salvation is that the believer will have peace in his heart when he is forgiven. The disciples had rowed about twenty-five or thirty furlongs. The only way a person can have peace in his life is to know that he is forgiven. It is a time of peace when Christ walks on the water and gets on board the boat with the disciples. Every believer who has experienced forgiveness has the peace of God concerning their future destiny. Forgiveness brings peace.

III. The Power of Forgiveness - Matthew 18:23

The theme of forgiveness is found throughout the Bible. Jesus tells the parable of a certain king (Matthew 18:23-35) who was taking account of his servants. The servant of the king receives forgiveness of a large debt. The king discovers later that this servant had a fellow servant which owed him an amount, and the servant refused to dispense forgiveness to his fellow servant. The king was angry and then demanded of the servant all that was due him because of the servant's unwillingness to forgive his fellow servant. Every person possesses the power to forgive others of their offenses.

CHAPTER 31

THE NUMBER TWENTY-SIX ~ "GOSPEL"

The Source

The number twenty-six is not found in the Bible. There are several hidden orders or lists in which the number twenty-six appears. The number does not appear, but the lists contains twenty-six items. In II Corinthians 4:3-4, Paul said,"If our gospel be hid, it is hid to them that are lost, and whom the God of this world has blinded the mind of them which believe not less the light of the glorious gospel of Christ who is the image of God should shine unto them."

The Symbolism

The number twenty-six symbolizes the Gospel or good news. From II Corinthians 4:3-4, one finds the Gospel is hid. The number twenty-six is hid and not easily found. It shines bright to the saved, but to the lost, the gospel is hid. The unsaved person has no idea of the Gospel. He knows he is a sinner and he should not be that way. He does not want to be that way, but there is pleasure in sin for a season. The seasons change and he finds himself in an awful shape. The Gospel is hid unto those that are lost. A lost man without the Word of God cannot come to Christ. It takes the Word of God and the Spirit of God working together for the sinner to be saved.

Noah's name is first mentioned in Genesis 5:29. The twenty-sixth time Noah's name is mentioned is in Genesis 8:11 when the good news is given to Noah that the waters were abated from the earth. It was good news to Noah and his family that they had survived the great flood by being inside the ark. They floated above the judgment upon the earth. Noah and his family went into the ark which was shaped like a coffin. Like dead people buried, they were not seen for days and they came out like a resurrection. It was good news when Noah knew the waters were abated from all the earth (Genesis 8:11).

The name of Moses is mentioned the twenty-sixth time in Exodus 4:19, where God gives Moses good news with these words, "Go return into Egypt for all the men are dead which sought thy life." In other words, there would not be revenge upon the life of Moses. Moses had received good news.

II Timothy 1:8 states, "To be thou partaker of the afflictions of the Gospel according to the power of God." In II Corinthians 11:23-27, it lists twenty-six things which Paul endured. Paul instructed Timothy to be a partaker of the Gospel that he had experienced. The good news is that a person can get saved, and God can help that him through every affliction and every problem.

The Sermon

The Gospel of Jesus Christ

I. The Definition of the Gospel ~ I Corinthians 15:2 - 3

John 3:16 is the gospel in concise words. "For God so loved the world that He gave His only begotten Son that whosoever believeth in Him should not perish but have everlasting life." In the English language, it has twenty-five words, but in the Greek language there are twenty-six words that make up this Gospel verse.

So what is the Gospel? In I Corinthians 15, Paul told the Corinthians that they had heard him preach that Christ died for their sins, that He was buried, and He arose again the third day according to the Scriptures. The good news is that Christ suffered and died according to the Scriptures for the sins of the world, and He was buried. He was dead because the wages of sin is death. He died on a cross. The good news is that on the third day He arose. The reason Christ died according to the Scriptures for the sins of the world was to save sinners. He did not just die to be a martyr. He did not hang on a tree so people would hold him up as some great religious leader. He died for the sins of the world according to the Scriptures. The good news is that three days after He was buried, He arose, and His tomb is empty.

II. The Light of the Gospel ~ II Corinthians 4:4

In Acts 9, Tabatha, who was also called Dorcus, had died. When Peter's name is mentioned three more times, which is the number of completion, Dorcus was resurrected or made complete from death. In Acts 9:40 is the twenty sixth time that Peter's name appears in the book of Acts, and Dorcus opened her eyes. The gospel will open the eyes of a listener.

III. The Gospel Is Hidden ~ II Corinthians 4:3 - 4

The number twenty-six is not found directly in the Bible. It is hid. The Gospel is hid to those that are lost. There are some lists in the Bible where the twenty-sixth time you will find the good news. Christ's death, burial, and resurrection is good news.

CHAPTER 32

THE NUMBER TWENTY-SEVEN ~ "HOPE"

The Source

The first mention of the number twenty-seven is found in Genesis 8:14 when Noah discovers that the earth has dried from the great flood. In the twenty and seventh year of Asa, King of Judah, Zimri reigned in the Northern Kingdom (I Kings 16:15). In I Kings 16:15-20, the evil king, Zimri, commits suicide. In II Kings 25:27-30, King Jehoiachin is released from prison by his conqueror, Evilmerodach, king of Babylon on the twenty-seventh day of the month. In Ezekiel 29:17 on the twenty-seventh day of the month, God spoke to the Prophet Ezekiel and told him the future concerning Nebuchadnezzar and the land of Egypt.

The Symbolism

The idea associated with the number twenty-seven in the Bible is hope. In Genesis 8:14, Noah has hope in a new life in a new world when he discovers the land is dried after the great flood. In I Kings 16:15-20, the evil king, Zimri, commits suicide which gave hope to the nation of Israel because he was so evil. In II Kings 25:27-30, King Jehoiachin is released from prison which gives hope in his life. King Jehoiachin probably expected execution at the hands of Evilmerodach.

In Ezekiel 29:17 the number twenty-seven is found for the fourth time and last time in the Bible. God commanded Nebuchadnezzar to battle an old enemy and take over Tyrus. Nebuchadnezzar had suffered financial lost and received nothing from the war because Tyrus burned and was destroyed before he arrived. God told Nebuchadnezzar He would give him the land of Egypt as wages for his army. God told him to take the spoils of the land and heathen people for his wages. "Because they wrought for me," saith the Lord God. There is hope as evil empires are working for God and Israel buds forth. God is recognized as the ruler of this earth.

The twenty-seventh time the word *hope* is found is in the New Testament in I Corinthians 9:10. The word hope, in this one verse, appears three times. It is rare that the word would appear in one book, but here is hope appearing three times in one verse. Three symbolizes completeness. The middle time it occurs is when it appears the twenty-seventh time. The verse says, "Or saith He it altogether for our sakes? For our sakes, no doubt, this is written: that he that ploweth should plow in hope." The farmer who puts his hands to the plow and lays off the furrows of rows does it with an expectation of harvest. He does it with hope that a crop will come up where he plows. "That he that ploweth should plow in hope, and he that thresheth in hope should be partaker of his hope." The thresher is the man who cuts the grain and ties it in bundles for the harvest. That is the twenty-seventh time that hope is mentioned. Hope has its fruit in the harvest that will come.

The number twenty-six is symbolic of the gospel which is hidden. We see the suffering for the Gospel which the Apostle Paul endured. Paul suffered twenty-six different incidents in the furtherance of the Gospel. It was hidden to man. The number twenty-seven is hope. The number twenty-seven appears four times in the Bible, which is the number of mankind. There is hope for man.

In mathematics the number twenty-seven is representative of the cube. Twenty-seven is the product of three times three times three ($27=3 \times 3 \times 3$ or 3^3). Twenty-seven is symbolic of hope or completeness threefold.

The Sermon

Our Blessed Hope in Christ

I. Our Hope in Christ ~ Genesis 8:14

We have hope in Jesus Christ. As Noah exited the Ark, he started a new life. Everything he possessed had been destroyed except what was aboard the Ark. The sinner that comes to Christ has a new life in Christ. Old things are passed away, and all things are become new. Things will look different when a person puts his faith and trust in the Lord Jesus Christ. Noah stepped off the ark and did not recognize any thing because the water had obliterated it. When somebody comes to Christ, it is with expectation. It is expecting hope that there is a Savior who walks with him. Just as Noah had a new life, the sinner who comes to Christ has hope of a new life.

II. The Doom Without Christ ~ I Kings 16:20

In I Kings 16:20, Zimri was taken off the throne. He arose by a conspiracy, took the throne, and ruled over Tizrah. In horror of what had happened, the men of Israel told Omri to be their king. Omri led a force into Tizrah, and when they advanced, Zimri saw the handwriting on the wall. He ran to the king's house, set it on fire, and burned it all around him. He burned himself to death inside that ring of fire.

Jehoiachin was captured by Evilmerodach, the king of Babylon. Evilmerodach displayed something that is rarely displayed by rulers. He called to the prison and instructed the warden to bring out Jehoiachin. Evilmerodach brought the king out of prison, asked him to put on his kingly robes, spoke to him kindly, gave him an allowance, and said he could eat in his court on a daily basis the rest of his life. That is a picture of salvation. Here is a second chance at life after being released from the prison of sin.

III. Our Hope is in a Kind and Good God ~ Ezekiel 29:17

It was God who led Nebuchadnezzar, because he did not get any pay in Tyrus, to conquer Egypt for wages. God stated that He is the one ruler of the world, and he, Nebuchadnezzar, is working for Him. God is a kind, loving, and long suffering God. He is a great God to serve. For the sinner, here is a hope of a new life in Christ and a second chance at life. Even in prison, there can be a second chance. When the Gospel is given and men believe in Christ, there is a change in their life and there is hope.

In days past, when a young lady was around twelve years of age, her parents would purchase for her a cedar chest. In that hope chest, the young lady would put plates, cups, silverware, and things that she would need in her new home when she was married. A coffin looks like a hope chest. Believers have the hope of the Word of God that the body that has been buried will one day come to life and come out of the grave. That is the greatest hope chest any family member can have when they make the trip to the funeral home to make arrangements for a dead family member. The believer of Jesus Christ has a blessed hope of the appearing of the Lord Jesus Christ.

CHAPTER 33

THE NUMBER TWENTY-EIGHT ~ "ATONEMENT"

The Source

The number twenty-eight appears in our Bible only three times. It does not appear in the New Testament. The first mention is in Exodus 26:2 where the pattern of the Tabernacle is given to Moses. God said to make ten curtains. He specified the material to use, and gave the dimensions of the length of one curtain, twenty-eight cubits. One cubit is the measure from the High Priest's longest finger to his elbow. The High Priest came from the lineage of Aaron, and they were not permitted to marry outside their own tribe. Because of the specification of marrying within their own tribe, the next high priest was just about the same size of the high priest before him. The standard measurement was about the same from generation to generation in the tribe of Levi. A cubit is approximately sixteen to eighteen inches.

The cubit gave the children of Israel a unit of measurement, and God said that the length of one of these curtains will be twenty-eight cubits, and the breadth will be four cubits. It was not tall, but it was long. It was seven times as long as it was tall. There are four numbers mentioned in these verses. He said to make ten curtains. Ten is symbolic of order, and God works in an organized manner. God has an order in what He is doing in the lives of believers with respect to numbers and the Scriptures. God has not put one useless word in the Bible. God has not omitted any words necessary in his Bible. Every word in the Bible appears for a purpose.

In II Kings 10:36 the number twenty-eight is mentioned. Jehu's reign over Israel lasted twenty eight years. In II Chronicles 11:21, Rehoboam had twenty eight sons and sixty daughters.

The Symbolism

The number twenty-eight symbolizes atonement. It is found predominantly with the Tabernacle. The word atonement means a covering. The linen cloth or linen curtain, which was twenty-eight cubits long and four cubits high, provided a covering for the sides of the tabernacle. It did not cover the top. Badger skins were stretched over the top. The number four is found here, for it is four cubits in the breadth. Four is symbolic of the world and every person of the world is included in God's plan. God has a plan for every boy, girl, man, and woman in His grace. God said to hook five curtains together. The other five were coupled to be used so that the priests could put one curtain at the front and extend four curtains along the side. The other side would be constructed in the same manner. Five is the number of grace, and this was the number of curtains which were linked. Titus 3:4 gives us the definition of grace, which is the kindness and love of God toward man appeared. There is grace which God has provided with an atonement for every person.

The Sermon
I. The Atonement in the Old Testament ~ Genesis 3:21

God always does things in an organized manner. In the garden of Eden, God took the skin of animals and made coats for Adam and Eve. The type of animal that God required was a lamb in later sacrifices. The lamb was used as the Passover sacrifice. The lamb was used as a morning sacrifice and evening sacrifice. When John the Baptist saw Jesus, he did not say for his disciples to behold the bull, or behold the goat, or behold the pigeon, which were used in sacrifices, but he said, "Behold the Lamb." It is a good probability that God in the Garden of Eden used a lamb to make the coats for Adam and Eve. Once an animal had been killed and skinned, there remains blood on the inside of the skin. When God put coats on Adam and Eve, there were possibly spots of blood on the skins of Adam and Eve. If one removed those coats from Adam and Even, and looked closely, a blood covering would be seen. The atonement says that when God sees the blood, He will pass over it. God saw it over the doorpost of the Jewish people living in Egypt, and knew they had made the sacrifice that would please Him. When the number twenty-eight is given in the book of Exodus, it is an indication that a covering was present. It is the covering for the tabernacle.

II. The Atonement in the New Testament ~ Romans 5:11

Romans 5:11 says that we joy in God through our Lord Jesus Christ by whom we have now received the atonement. For linen to be made, the flax had to be beaten as Christ was beaten and scourged. The flax was whipped, beaten, and then it became pliable. The flax was then able to be woven into linen. Christ was beaten, and He became our atonement. The covering for Israel's sins consisted of a lamb or an animal being slain, and the blood would be the covering. It was not necessarily the skin; it was what was inside that skin that made the atonement. The blood was a covering for Israel's sins, and Christ's blood is a covering today for all of man's sins. When Christ died on a cross, He died for every single person on the earth. To receive the atonement, a person has to call upon God. It is a gift offered to man, and a person must receive that gift. Romans 5:11 declares that Christ is the atonement for sins.

III. The Atonement and the Grace of God ~ Titus 2:11

The grace of God points in every direction around the Tabernacle. The linen curtain walls were connected with five curtains so they could transport it easier. The number five is symbolic of the grace of God. God gave grace to the world through His Son. His kindness and love was demonstrated to the world through the gospel of Christ. Titus says that the grace of God that brings salvation has appeared unto all men (Titus 2:11). God, through His grace, gave us His Son. He was the atonement for the sins of the world. He was the covering for the sins of man.

CHAPTER 34

THE NUMBER TWENTY-NINE ~ "SANCTIFICATION"

The Source

The first mention of the number twenty-nine is in Genesis 11: 24. Nahor is the grandfather of Abram, and the word Nahor comes from the Hebrew word which means "snort, snore, or nostril." The nostrils, were the location of the Adam's body, where God breathed life into Adam. Nahor was destined to bear Terah, who was to bear Abraham, the father of faith. It is interesting when he mentions the number twenty-nine, he mentions nostrils. After God had created the features of man from the dust of the earth, man did not have life. God could have spoken life into Adam, but He chose to breathe into his nostrils the breath of life.

The number twenty-nine is again used in Exodus 38:24, which says that all the gold that was needed for the work of the holy place, even the gold of the offering, was twenty-nine talents and seven hundred thirty shekels of the sanctuary. There were twenty-nine talents of gold in the holy place. This was given to the Temple to be used in a holy way. It was sanctified. It was set aside for a purpose. A sanctuary is a place that has been set aside for God. The gold given in Exodus for this offering was set aside to do the work in the holy place.

In Joshua 15 there are twenty-nine cities with their villages. The tribe of Judah had twenty-nine cities. God had chosen to bring forth the Messiah through the tribe of Judah.

In Ezra 1:9, there is an interesting story using the number twenty-nine. The verse shows there were thirty chargers of gold, a thousand chargers of silver, and twenty-nine knives. These were not knives that men carried with them for everyday use. These were the sanctified knives used in the animal sacrifices for the Temple. These twenty-nine knives were used in the daily sacrificial ritual. They were the vessels from the house of the Lord which Nebuchadnezzar had brought forth out of Jerusalem. He had put them in storage, and when the time came for Ezra to enter Jerusalem and rebuild theTemple, these would be used for the animal sacrifice.

The Symbolism

The number twenty-nine symbolizes sanctification or to be set aside to do a work for God. Sanctification has two parts which include being separated from the world and being set aside for a purpose. The believer is to be sanctified in the work of the Lord. In Genesis 11:24, Nahor whose name means "nostril," was set aside by God to bring forth Terah who was to birth Abraham, the father of faith.

In Exodus 38:24, the gold offering for the work in the holy place was to be twenty-nine talents. In Joshua 15:32, the tribe of Judah has twenty-nine cities noted. From the tribe of Judah would be born the Messiah.

The knives mentioned in Ezra 1:9 symbolize a cutting away from the world or sanctification. The word sanctification is associated with the word "saint." Peter called the believers "saints."

The number twenty-nine is a prime number which means it is divisible by itself and one. No other number will evenly divide into twenty-nine. It has a twin prime number, which is the number thirty-one. The numbers twenty-nine and thirty-one make up the sixth pair of prime twin numbers. It is interesting that sanctification is a Biblical principle that applies mostly to man, which is the number six. Although a building or piece of property can be sanctified, God would rather man be sanctified than a piece of property or a building.

The Sermon

Sanctification

I. The Meaning of Sanctification ~ I Corinthians 6:11

Sanctification is being set apart from the world with a purpose for God. The gold which was given for the Temple was set apart to be used in it. The knives used to prepare the sacrifice were set apart for a specific purpose. The believer is sanctified through the Word of God and through the Spirit of God.

II. The Method of Sanctification ~ John 17:7

John 17:7 tells how a believer becomes sanctified. When a person reads the Word of God, or hears the Word of God, that person has knowledge of how to conduct himself. As the Word of God works in the life of a believer, it is like a sword that cuts off things that are not right in his life and sanctifies the believer. Peter said in I Peter 1:2 that we are sanctified through the Holy Spirit. The Holy Spirit is every believer's teacher who is living in the believer and walking with the believer.

III. The Separation from the World ~ II Corinthians 6:17

The number twenty-nine tells a story of how the Jewish people sanctified the gold in Exodus 38:24. God sanctified the lineage of Abraham from Nahor in Genesis 11:24. The Old Testament shows the believer examples of how God had set aside things to be used in the Tabernacle and Temple. God set aside not only articles, but He also set aside people to be used by Him. God wants His people to be separated from the world, and to be used by Him for His purpose.

CHAPTER 35

THE NUMBER THIRTY ~ "THE LIFE (BLOOD) OF CHRIST"

The Source

The first mention of the number thirty is in Genesis 6:15. God gave the dimensions of the ark. The wooden ark was to be three hundred cubits by fifty cubits and have a height of thirty cubits.

The number thirty is found in Matthew 26:15 when Judas Iscariot bargained with the chief priests for the deliverance of Jesus to them for the amount of thirty pieces of silver. In Matthew 27:3-8, Judas Iscariot returned to the Temple after betraying Jesus in the Garden of Gethsemane and cast down the thirty pieces of silver. The chief priests declared it to be blood money and refused to put it into the Temple treasury. They took counsel and bought a potter's field. A potter's field is where the potter would discard his broken, useless pottery. The field was called The Field of Blood. The potter's field would be used to bury strangers or the very poor.

In Genesis 41:46, Joseph was thirty years of age. In Numbers 4:3, the high priest was to be thirty years of age and no older than fifty years of age. In II Samuel 5:4, David was thirty when he began to reign, and he reigned for forty years. David became thirty years of age when he began his reign over the kingdom in Judea. In Luke 3:23, Jesus was about the age of thirty when He began his earthly ministry.

The Symbolism

The first mention of the number thirty is found in the dimensions of the wooden ark which God had commanded Noah to build. The wooden ark was to be three hundred cubits by fifty cubits and have a height of thirty cubits. The symbolism of the number three hundred is the number of the Father. Fifty is symbolic of the Holy Spirit. The word Pentecost means fifty days, and this was the time when the Holy Spirit first came to believers . The number thirty is symbolic of the Son of God; the life of the Son of God, the blood. Noah survived God's judgment upon the world because of the Ark whose dimensions represent the Father, the Holy Spirit and the Son. If you look at the dimensions, 300, 30, and 50, they are proportional to a casket. If a person viewed the ark from the heavens, at God's point of view, he would see an object like a coffin floating through the waves and over the oceans. This is a picture of the resurrection. Noah and his family went into the Ark and were shut in by God. A year and seventeen days later, they came out of the Ark and multiplied and replenished the earth. Seventeen is symbolic of victory in the Scriptures.

The number thirty has a significance of coming to age in the Bible. The number thirty seems to be the age of maturity. It is the age when God and society places responsibility upon a person. In Genesis 41:46, Joseph was thirty years of age when he became second in command in Egypt. He had given Pharaoh the interpretation of his dream. Before he came out of prison, Joseph did one thing. It is the first time it is found in the Bible. He shaved. Joseph was old enough to grow a beard or mustache. He shaved because the Egyptians did not like beards or growth of hair on their faces. When he came out, he respected Pharaoh by knowing and following the customs of the land. Joseph had been tried by God and proved himself faithful at the age of thirty, when he came out of the prison and ascended to the throne as the ruler behind Pharaoh.

In Numbers 4:3, the high priests were to be of thirty years of age and no older than fifty years of age. They had a twenty year span they could minister as high priests for the temple. Most priests were eligible to be the high priest.

In Luke 3:23, Jesus was about the age of thirty when he began his earthly ministry. When he went to the marriage of Canaan, his mother informed Him that the grape juice refreshments were depleted. Jesus replied that His time had not yet come. The miracle of changing water into wine was to show the disciples who He was. They had already heard John the Baptist say, "Behold the Lamb of God that taketh away the sin of the world." This miracle proved to the disciples that Jesus was from God.

The Sermon
The Blood of Jesus Christ

I. The Blood Sacrifice before the Law - Genesis 9:4 - 5

Before God gave Moses directions for the blood sacrifices in the book of Leviticus, He instructed Adam on a blood sacrifice for the remission of sins. This is evident in the fact that Abel brought an animal to be sacrificed instead of vegetables from his garden. By word of mouth, the descendants of Adam were taught the correct manner to offer a blood sacrifice.

II. The Blood During the Law - Leviticus 17:11

When God instituted worship at the Tabernacle, the central element was the blood sacrifice. God gave Moses instructions on the correct manner for the descendants of Aaron to follow. The blood sacrifices were specified for different transgressions by the children of Israel. The blood flowed every day from the knives of the priests as they obeyed God in offering their sacrifices.

III. The Life of Jesus - John 6:53-54

In the New Testament, Luke 3:23 states, "Jesus himself began to be about thirty years of age", when he began his public ministry. Although Jesus did many great deeds, His ministry was to come to earth; be the Lamb of God; and die on a cross for the sins of the world. He was buried and rose from the dead on the third day. His ministry was to give His life, His blood, for lost sinners. The Lord Jesus Christ came to seek and to save that which was lost. He started His ministry when He was of the age of thirty.

Jesus came to earth as the Lamb of God to take away the sin of the world. Judas Iscariot betrayed Jesus for thirty pieces of silver which was eventually used to buy the field of blood. The number thirty symbolizes the life of Christ, and life is synonymous in the Word of God with blood (Leviticus 17:11).

CHAPTER 36
THE NUMBER THIRTY-ONE ~ "CLEANSING"
The Source

The first mention of the number 31 is found in Joshua 12:24, where King of Tirzah was one of thirty-one kings. In Joshua 12:8, Joshua had been given the authority to cleanse the promised land of different tribes of people. Then Joshua starts naming 31 kings. In verse 24, he names the 31st king and then says, "all the kings thirty and one." The number thirty-one is only mentioned three times in the Bible.

Josiah became king when he was 8 years old, and he reigned thirty and one years in Jerusalem (II Kings 22:1-2). He had many governmental advisors and people around him to assist him. Of the kings that followed King David, Josiah was a great ruler. He ruled thirty-one years.

The life of King Josiah is also given in II Chronicles 34:1. Now it is a little strange here, because instead of thirty and one years, the Bible says one and thirty years. It is the same story in Kings, but it reveals a little more. An eight year old boy just became a king and started seeking after God. That is not the normal, rather it is unusual. In the twelfth year, he began to purge Judah. So eight plus twelve makes him twenty years of age. He began to purge Judah by cutting down the idols of

Balaam. King Josiah ground up the idols; went to the graveyard and strew the ashes over the cemetery like grave yard dust.

The Symbolism

The number thirty-one is found only three times in the Bible. The number thirty-one symbolizes cleansing. In Joshua it becomes clear that God was cleansing the promised land of the unclean kings in their kingdoms. Joshua conquered the land; captured the enemy kings or did away with them; and dethroned them. One of the first things mentioned about Josiah when he turned thirty-one was that he cleansed the religious crowd. All the worshipers of Balaam and all their graven images which were against the Word of God, were taken by King Josiah who ground them into dust, and scattered them over the graveyard. Josiah did a cleansing of the land in removing those kings and kingdoms so the righteous could occupy the land. Josiah cleansed the land also by taking out the religious realm.

There are some mathematical comparisons with the number twenty-nine and the number thirty-one. The number twenty-nine is sanctification. Number thirty is the life or blood of Christ. The number thirty-one is symbolic of cleansing. The numbers twenty-nine and thirty-one are twin prime numbers just as sanctification and cleansing are similar terms. There are very few twin prime numbers in the number system. They are both primes, which means the only number that will divide evenly into them is one and the number itself. The numbers two and three are the first twin prime numbers although they are not separated by one number. The next twin primes are the numbers three and five; then five and seven; then eleven and thirteen; then nineteen and twenty-one. The next set of twin prime numbers is twenty-nine and thirty-one.

The Sermon

The Cleansing

I. The Place of Cleansing ~ Exodus 30:18

The place used specifically for cleansing was the laver. The laver was the brass bowl in which water was placed to cleanse the priest before he could do any service in the Tabernacle or Temple. Exodus 30:18 tells where Moses was given the directive from God to build the laver, and it was a type of judgment made out of brass. Brass is the symbol of judgment. Water symbolizes the Word of God which is the judge. Who is the believer's judge? Believers do not have to judge one another. It is the Word of God that judges the believer. The believer can go to the judgment seat of the Word of God and it will tell him whether he is right or wrong. The Author of the

Bible who lives in the believer will convict the believer of the deeds that are wrong.

II. The Particles of Cleansing

Most people think that washing detergent cleans their clothes. Scientists state that the soap simply breaks down the water molecules so fine that the water passes between the fibers of the soiled garment. The soap does not perform the act of cleansing. It is the water that cleans the clothes. In the Midwest, farmers spray a soap solution upon their crops just before a rain. When it rains, the water will break down into small molecules and will drain into the soil to water the plants instead of running off the fields. Water has a magnetic characteristic that allows it to cling to itself. This is a great act of the Creator who created the water molecules with a bonding property. The water or sap in a plant goes from the roots to the leaves without the plant having a "heart" to pump it. When a water molecule evaporates from the leaf of a plant, it pulls the next water molecule into place. God is a magnificent Creator.

The laver in the tabernacle had no determined size given in Exodus 31. No one knows the depth, nor the scope, nor the value of the Word of God. The Word of God is our laver. We are not told how the priests transported the laver. There is no description in the word of God how it was to be transported with the tabernacle. In the temple it remained stationary, and Solomon built several lavers what were used by the priests as they prepared for the great number of sacrifices.

The Word of God is the water that washes through the fabric of our lives and cleanses us. For some believers it only takes a small amount of the Word of God to cleanse them, while others take a large amount of the Word of God to purify them. In either case, it is the Word of God that is the element used to accomplish the cleansing and purging.

III. THE Person of Cleansing

Speaking of the Church in Ephesians 5:26-27, Paul stated, "That He might sanctify and cleanse it with the washing of water by the Word, That He might present it to Himself a glorious church, not having spot, or wrinkle, or any such thing; but that it should be holy and without blemish." God not only uses the Word of God to cleanse the believer, but the Holy Spirit is the key Person in this cleansing. When the believer reads the Word of God, it washes and cleanses him. The Holy Spirit is like the soap agent that reveals the Word of God to the believer so the believer can see the areas of his life that need cleaning.

A grandfather was attempting to teach his grandson a lesson. The grandfather gave the grandson a basket and sent him to the spring to get a basket full of water. The little boy ran to the spring, dipped the basket in the water, pulled it out, and ran as fast as he could back home. By the time the grandson got back to the house the water had seeped out of the basket. On the third attempt to bring water to the grandfather from the spring, the little boy told the grandfather that getting water from the spring with a basket was not a good idea. The grandfather with his wise voice said, "Yes, it did. Look at your basket. The basket is real clean right now." Sometimes when the believer reads the Word of God, it seems in vain, but the Holy Spirit can take the water of the Word of God going through the believer to wash, cleanse, and keep him clean from the things of the world. James 4:8 gives the believer a commandment from God to "Draw nigh to God, and he will draw nigh to you. Cleanse your hands, ye sinners; and purify your hearts, ye double minded." The believer is told to clean his hands. He is told to purify his heart. There is a command to cleanse so the believer may draw nigh to God, and He then will draw nigh unto the believer. First John 1:9 states, "If we confess our sins, He is faithful and just to forgive us our sins, and to cleanse us from all unrighteousness." If the believer willfully, or by accident sins, then there is a recourse and a remedy. The anwer is confessing sins, agreeing with God that it was wrong, and then

God is faithful and just to forgive the believer of his sins. The Word of God and the Holy Spirit, will do the cleansing.

Sanctifying and cleansing are not the same. If it had been the same Paul would have used just one word, but he said, "sanctify and cleanse" in Ephesians 5:26. In Ephesians 5: 26, the twin prime words "sanctify" and "cleanse" are both found. The number twenty-nine is symbolic of sanctification. The number thirty-one is symbolic of cleansing. The numbers twenty-nine and thirty-one are twin prime numbers which walk together.

Sanctification is the initial and ongoing process to keep doing right. Sanctification occurs the moment we get saved and it is progressive. It does two things: (1) separates us from the world; and (2) separates us for service. The believer is a child of God and has a purpose in life. Sanctification is for the believer to keep doing right, living right, and staying away from the world.

Cleansing is a daily process to stop doing evil. Paul said in I Corinthians 15:31, in effect, that, "I die daily". He said that as he walked along this road, he picked up dirt. As the believer associates with the world, he is going to need cleansing. The believer cannot live in this sinful world without it affecting him. The believer must die daily as Paul said in I Corinthians chapter fifteen. Lava Soap is a common brand of soap. In the Spanish language the word "lava" means "you wash." Most believers do not deliberately pick up dirty things in the world but it happens. Keeping clean between the ears may be more important than keeping clean behind the ears. Being in the world puts a spot or a blemish on the believer that he doesn't even realize. The proper way for the believer to cleanse himself is by reading the Word of God. It gets the believer cleaner every time It is read. It is like water going through the dirty basket. When the believer listens to the Holy Spirit in prayer, then God will bring to the remembrance of the believer his transgressions. The believer can then confess his sins and be cleansed.

CHAPTER 37
THE NUMBER THIRTY-TWO ~ "COVENANT"

The Source

The number thirty-two is found in II Chronicles 21:5-7, but it is first mentioned in Genesis 11:20 where "Reu lived two and thirty years, and begat Serug." According to the Law of First Mention the text of Genesis 11:20 should be examined to see if the meaning is true in the other times the number thirty-two is mentioned in the Bible. The Law of First Mention is always a good rule of thumb, although it is not a solid rule. In Numbers 31:40, there are thirty-two persons who are the Lord's tribute. In II Chronicles 21:5-7, the story of Jehoram as King of Israel is related. He was thirty and two years old when he assumed the throne.

The Symbolism

The number thirty-two symbolizes a covenant. A covenant is defined as an agreement, usually a formal one, between two or more persons to do or not to do something specified. A covenant is simply an agreement between two parties.

Genesis 11:20 is a very interesting verse. It is the first time the word thirty-two is mentioned in the Bible. The name *Reu* means "friend" in the Hebrew, and the name *Serug* means "tendril or a climbing clinging plant." Combined together those two names yield a covenant as being an agreement between "clinging friends."

The next time the number thirty-two is mentioned is in Numbers 31:40 where thirty-two persons who were captured in a battle against the Midianites. The children of Israel had killed all the men. The women who remained had never been married. The persons that are mentioned in verse 40 were actually given as a tribute to God. They were given, we think, to the workers and priests in the tabernacle, and these women were to help them in doing any tasks at the tabernacle. They were separated servants or slaves, and they were put in servitude serving in the work of the tabernacle. The covenant or agreement in this verse is the fact that the children of Israel had made a covenant unto God that whenever they captured their enemies, they would give God a tribute. They had made a covenant with God to do this, and these thirty-two persons were a tribute until the Lord.

In I Kings 20:1 a covenant is made between the king of Syria and thirty-two kings to go to war against Samaria. I Kings 20:1 states, "And Benhadad the king of Syria gathered all his host together: and there were thirty and two kings with him, and horses, and chariots; and he went up and besieged Samaria, and warred against it." Here is a war covenant.

Benhadad and thirty-two kings made an agreement and fought a war.

II Kings 8:17-19 tells the story of Jehoram and states,"Thirty and two years old was he when he began to reign; and he reigned eight years in Jerusalem. And he walked in the way of the kings of Israel, as did the house of Ahab: for the daughter of Ahab was his wife: and he did evil in the sight of the LORD. Yet the LORD would not destroy Judah for David his servant's sake, as he promised him to always give him a light, and to his children."

Jehoram, being thirty-two years of age, lived a wicked life; but God did not destroy the house of David because of the covenant that He had made with David. The action of the believer does affect lives of the little ones in their household. The believer can be guaranteed one thing, if God ever gives him an agreement, then God will keep his agreement with him. God made an agreement or covenant with David because of the way he had lived. God said that David's house will live forever and will reign forever. God did not put any conditions on it. He did not use the word "unless." He just said it would live forever. God endured some bad people in places of authority simply because God had made an agreement with David to keep the covenant He had made with him. God is not going to break his covenant. God has made an agreement in the Word of God. God will not break His Word. If God breaks his Word, He has lied and no more is He God. God cannot lie. Someone has said that God can do everything, but this is not true. God cannot die. God cannot lie. God cannot let anyone go to heaven who does not accept his Son as his savior. There are a lot of things God cannot do because He is God.

The number thirty-two appears about five times in the Old Testament. All around the number, you can see the mention of covenant or agreement that God has made with someone or a nation. The number thirty-two is not found in the New Testament, but covenants are mentioned.

The mathematical analysis is unique. Thirty-two is a five time multiple of the number 2, or 2x2x2x2x2 which equals 32, or in mathematical terms, 2^5. Many numbers can be formed using the divisor of 32. Thirty-two has one of the largest number of divisors, numbers that will go into it evenly: 1, 2, 4, 8, 16 and 32. A covenant can be between just two person, or a covenant can be between many. The church has a covenant, that being joined together we agree to these things. In the covenant between David and God, God made a covenant with just one person. God made a covenant with Abraham and He included all his family, an innumerable host.

Sermon

I. God's Covenant with Adam ~ Genesis 2:15-17

In Genesis 2:15-17, God made a covenant with man. It was the covenant of life. A covenant has rules. God agreed with Adam for him to live in the new creation, but if in the day that he ate of the forbidden fruit he would surely die. Forbidden fruit has gotten many men into a jam. Adam did not physically die that day, but he spiritually died that very hour, and the nature of sin and the consequence of death was passed on to every descendent. The proof that humans were born of Adam and Eve is the fact that everyone dies. The covenant was broken by man.

Another phase of the Adamic Covenant is found in Genesis 3:15 where it states, "And I will put enmity between thee and the woman, and between thy seed and her seed; it shall bruise thy head, and thou shalt bruise his heel." This was the covenant given to Adam, Eve and all mankind stating that there was going to be One to come that will set the believer free. He will give by the seed of woman the One who will conquer the enemy. The Lord Jesus bruised the head of the devil, and the devil bruised His heel. Satan's head was bruised at Calvary, and the heels of Jesus were bruised by the Roman soldiers as they nailed Him to the cross. The Adamic Covenant was fulfilled in the New Testament by the Son of God laying down His life on the cross.

II. God's Covenant with Abraham ~ Genesis 12:1-3

God's request to Abraham was simple. Leave the land of your birth and go where I send you. The promise was simple. Obey me and I will bless you. The blessings extended to all the children of Israel if they obeyed God. The blessings also extended unto all the families of the earth.

In Exodus 19:5-6 God repeats His Covenant with Abraham and his descendants by saying, "Now therefore, if ye will obey my voice indeed, and keep my covenant, then ye shall be a peculiar treasure unto me above all people: for all the earth is mine: And ye shall be unto me a kingdom of priests, and a holy nation. These are the words which thou shalt speak unto the children of Israel." God made a covenant with the children of Israel that they would be a peculiar treasure above all people, because all the earth is God's. God entered into a special agreement with the children of Israel.

III. God's Covenant with Man ~ Hebrews 8:8

God has a new covenant not only with the house of Israel and with the house of Judah, but with any person who is willing to repent of their sins and be born again. It is not a covenant like the old covenant. In the New Covenant, God has promised to put His laws into their mind, and write them on their hearts. God has promised that He will be a God to them and they shall be a people to Him.

The word "testament" is the same as "covenant." The entire Bible is a covenant from God to man, an agreement with man. In the New Testament, God has so many promises. The wonderful promises of God are in God's new covenant, the Bible. The believer needs to take it, believe it, and walk by it. God knows what is good for his children, and He knows how to give good gifts to his children. The believer has a heavenly Father and a covenant with God.

CHAPTER 38

THE NUMBER THIRTY-THREE ~ "PURIFICATION"

The Source

The number thirty-three is found just a few times in the word of God. The first mention of the number thirty-three is found in Genesis 46:15, which was the number of sons and daughters of Jacob. In Leviticus 12:4, the days of purification of a mother after having a male child was thirty-three days. If she had a female child, then she would continue in the blood of her purifying sixty-six days.

The Symbolism

The number thirty-three symbolizes purification. The number thirty-three in the Old Testament is part of the Levitical law of purification for women after having male children. The number thirty-three stands for purification or cleansing internally. After a woman's body would deliver the child, the womb would be cleaned out, and her body would enter a cleansing stage. It would take eight days plus thirty-three days, or forty-one days after the child was delivered that she would be cleansed.

King David is mentioned in association with the number thirty-three. It is very unusual in that it was his very first act and his very last act which spanned a period of thirty-three years. His first act was cleansing.

In II Samuel 5:6-9, King David cleansed the city of Jerusalem of the Jebusites of which many were lame and blind. The Jebusites were the enemies of God, and many of them were lame and blind. A lame person who is not right with God can be a dangerous person. A blind person can hear sensitive information and report it to other people. They were lurking enemies of whom many were lame and blind. David had the enemy killed. It may seem cruel to us, but if they remained in the city of Jerusalem, David would have constantly fought them. In his first act, David cleaned out Jerusalem.

In I Kings 2:1, you will find David's last act as King. He instructs Solomon to purify the holy land of Shimei, an enemy of God. David's last act as king was to charge his son Solomon to beware of this man called Shimei who was an enemy of God. David had preserved his life during his reign.

David was returning from battle with Abner and Joab. Shimei stood on a hill and called David a dog, threw rocks at him, and cast dirt on him. Shimei was no immediate threat to David but he was an enemy. David's very last act which was thirty-three years later was to tell his son, "Solomon, now of all the things you need to do, watch out for Shimei." One of the first acts that Solomon did as king was to have an audience with Shimei and told him that he must keep within the boundaries of the holy city and do not pass over the brook Kedron. One day the report came back to Solomon that Shimei had crossed over the brook Kedron but had returned to Jerusalem. Kedron was the brook into which the blood sacrifice from the Temple flowed. Solomon commanded Shimei unto his throne and questioned him about crossing the brook Kedron. Shimei responded that he had crossed the brook Kedron. Solomon did not speak with Shimei, but called the officer of the guard to take his head off and take his body outside. Shimei was killed on the spot in the presence of Solomon. That seems rather cruel, but there are some things which should be immediately cut out of the life of the believer.

The number thirty-three is not found in the New Testament. In mathematical analysis, the number twenty-eight which symbolizes atonement, added to the number five which symbolizes grace, equals thirty-three which symbolizes purification. When a believer is saved by the grace of God, the believer receives the atonement and grace of God in their life. This results in purification in the life of the believer. I John 3:3, states that the believer is to purify himself as God is pure. Peter states that we are pure even as He is pure. Peter says we should be holy as He is holy. The number thirty-three points us to the point of purification.

The Sermon

The Life of the Believer

I. Cleansing of the Believer ~ James 4:8

Cleansing is to rid ourselves on the outside of anything that is evil. Sanctification is to be set aside to do good for the cause of God. Cleansing gets rid of evil, and and sanctification prepares us for service for the Master. Neither of these is purification.

II. Purification of the Believer ~ James 4:8

Purification is the daily cleansing internally. An illustrations of a Christian man is appropriate for this point. For some reason the Christian man sees pornography, says it is wrong, and he puts it away – that is cleansing. Then he turns to his Bible and starts reading, that is sanctification. The image that he saw went through his eyes and into his brain. At an opportune time, the devil will take that pornographic image and bring it to his memory. Now the Christian man needs purifying, not cleansing. He has already put it away by laying down the pornographic literature. He has turned to the Bible, which is sanctification. Inside the Christian man though, is a problem because the pornographic image is burning in his mind. When the Christian man realizes it is burning in his mind, he needs to drop to his knees, repent and ask God to replace it with the Word of God. For purification, one of the best things that can be done by the believer is to find a good Bible verse, like Psalms 101:3 which says, "I will set no wicked thing before mine eyes", and hid it in his heart.

Cleansing is when the believer says it is wrong and puts it aside. Sanctification is when the believer does something good for the cause of Christ. Purification is when the believer realizes that his mind has been polluted by sin, and prays to God to purify it completely. Washing cleans the outside and purifying cleans the inside.

III. Sanctification of the Believer ~ John 17:17

Sanctification is to be set aside to do good for the cause of Christ. The believer can overcome evil by doing good things. Hiding the Word of God in the heart of the believer will prevent the believer from sinning against God. The devil's goal is to snare us and put something in us that defiles us.

In review, the number twenty-nine symbolizes sanctification. The number thirty-one symbolizes cleansing, and number thirty-three symbolizes purification. Here is a triplet in the Word of God that every believer should realize in their life. God sanctified us, but there are remaining things the believer must do. The believer must grow, become dedicated to God, and develop a deep relationship with God.

CHAPTER 39
THE NUMBER THIRTY-FOUR ~ "SPIRIT-FILLED"

The Source

Number thirty-four is a rare number in the Word of God. It is only found one time. In Genesis 11:16 the Bible says, "And Eber lived four and thirty years, and begat Peleg." Very little information is given about Eber and Peleg, but we know that Eber was thirty-four years of age when Peleg was born as his son. Their names, like most of the names in the Bible have a deep meaning. In the Word of God, the names of men and women are placed in the Holy Scriptures for a reason. God never simply wrote words to "fill up" a book. Jesus said every jot and every tittle was important. Every little point is there for a reason. Nowhere in the New Testament is the word thirty-four found. The number thirty-four is very rare and so is the symbolism of the number. Eber and Peleg were placed exactly in the Bible for a very obvious reason when the symbolism of the number thirty-four is revealed.

The Symbolism

The number thirty-four symbolizes being Spirit filled. It takes diligence to be Spirit filled. Like the number thirty-four, being Spirit-filled is very rare. For clarity, the believer receives the Holy Spirit at the moment of salvation. God puts the believer in His hand and no man can take him out, not even himself. Being Spirit filled is much different than accepting the Holy Spirit into your life. One is cceptance and the other is a total submission.

The number thirty-four in the Old Testament is found in this one passage. This passage is like a vein of gold in the earth. It is rich and worth mining. The Hebrew Lexicon yields that the name Eber means "straight." The straighter a believer is, the more likely that person will be filled with the Spirit of God. The more a believer goes straight, the better things become. The believer who allows his life to be tangled in the affairs of this life, is going the opposite direction from a Spirit filled life. Eber means straight. The believer needs a straight life.

A search of the Hebrew Lexicon reveals that the word Peleg means "river." Job 29:6 is one of the first places it occurs where the Bible says, "When I washed my steps with butter, and the rock poured me out rivers of oil." Olive oil is symbolic of the Holy Spirit throughout the Word of God. Never in the Word of God does the word oil refer to crude oil which we use for fuel. It is always olive oil. Job said that God had blessed him so much that his steps were washed with butter and that the rocks flowed out rivers of oil.

The next time in the Bible the word *river* is found is in Psalms 1:3, "And he shall be like a tree planted by the rivers of water, that bringeth forth his fruit in his season; his leaf also shall not wither; and whatsoever he doeth shall prosper." The Bible is speaking about somebody that is straight in his living for God. Eber, after 34 years, gave birth to Peleg. After thirty-four years of living straight, then comes the river. Being filled with the Spirit of God immediately after salvation is very rare. It takes time to accomplish a life of living straight.

The tree planted by the rivers of water mentioned in Psalms 1:3 didn't come up from seed. It was planted by the rivers of water. Before salvation, the sinner is living in a barren land where there was hardly any water, no nutrients, and nothing to sustain him. At the moment of salvation, God uproots the new believer out of the old miry clay and plants the new convert by the river of life. There is a Spirit filled connection in this verse of the Holy Scriptures.

In Psalms 46:4, "There is a river, the streams whereof shall make glad the city of God, the holy place of the tabernacles of the most High." History tells us that it was Hezekiah who was directed by God to build an underground waterway. When a waterway is built, it is usually built straight. Hezekiah piped water from the spring outside the city of Jerusalem into the city. He concealed it with rocks so the enemy never knew of its location. Flowing into the city of Jerusalem was a supply of fresh, spring water.

In II Kings 18, the Bible relates the historical conquest of the Holy Land by Sennacherib, king of Assyria. The first thing a general like Sennacherib would do to conquer a city, was to isolate it. All the food and water would be withheld from the people inside the city and the people inside would deplete their source of food and water. They would finally raise the white flag and surrender. The Israelites had stored up a large amount of grain. Sennacherib made sure there was no outside source of water going into Jerusalem. What Sennacherib did not realize was that Hezekiah had built an underground aqua duct which brought fresh water to the residents of Jerusalem. They had plenty of water to drink and with which to cook. Sennacherib could not stop the flow of water because he did not know the source of the water.

There is a stream flowing from heaven which is the Holy Spirit. The Holy Spirit lives in us like a stream of fresh, spring water that the enemy cannot locate or stop. It is a stream from God. In Psalms 65, we see several references to the word Peleg, or river. The Bible says in Psalm 65:9, "Thou visitest the earth, and (Peleg) waterest it: thou greatly enrichest it with the

(Peleg) river of God, which is full of water: thou preparest them corn, when thou hast so provided for it." God visited the earth, watered, and enriched it with the river of God.

Eber means straight, and Peleg means river. In Genesis 10:25 the Bible says, "And unto Eber were born two sons: the name of one was Peleg; for in his days was the earth divided; and his brother's name was Joktan." Also in I Chronicles 1:19 the Scriptures state, "And unto Eber were born two sons: the name of the one was Peleg; because in his days the earth was divided: and his brother's name was Joktan." The fact of the earth being divided was a result of the waters receding from the earth after the flood in Noah's day. The continents were formed as a result of the water filling the oceans. Spiritually, there is a separation in the life of the believer which leads to a Spirit filled life. Separation from the world is very important. This is the idea of straightness, the dividing or separating from the world. The believer is not going to be Spirit filled if he is still part of the world.

Mathematical analysis of the number thirty-four is thought-provoking. It is the product of 2 x 17. The number two symbolizes witness. Seventeen is symbolic of victory. Therefore thirty-four is like a double victory. How could anything be a double victory? When the body of the believer is under the control of God and the spirit of the believer is under the control of God, then the believer has a double victory. Every person has three parts, the body, the soul, and the spirit, but every person has two natures. There is a carnal nature and a spiritual nature. People who are not saved only have one nature, the carnal nature. They cannot understand the Bible or spiritual things. They can only understand spiritual things when the Holy Spirit speaks to them. The number thirty-four is symbolic of being Spirit filled.

The Sermon

Be Filled With the Holy Spirit

I. The Necessity of Being Emptied - Ephesians 5:18

In Ephesians 5:18 the Bible says, "And be not drunk with wine, wherein is excess; but be filled with the Spirit". To be filled with the Spirit of God, one must first empty himself of the world. Is the glass half full or half empty? One can say the glass is half full of water and half full of air. If the glass is full of water, one cannot put grape juice in the glass until the water is first poured out. The believer cannot be filled with the Spirit of God if he has the world in him. The believer must first empty himself, and a good place to empty himself is at an altar of prayer.

II. Plant Your Life Around Christ ~ Psalms 1:3

In Psalms 1:3 the Bible says, "And he shall be like a tree planted by the rivers of water, that bringeth forth his fruit in his season; his leaf also shall not wither; and whatsoever he doeth shall prosper." When the believer is planted by the rivers of water, his life will be changed. It is not easy at times to be filled with the Spirit. Sometimes the believer gets in the flesh, and has regrets. When the believer plants his life around the Lord Jesus Christ at the river of the Holy Spirit, then it will not be long until that believer is filled with the Spirit of God.

III. Ask God to Fill You ~ Ephesians 5:18

The believer who has walked a straight life should ask God to fill him with the Holy Spirit. The believer should not dwell on that filling, rather have the Spirit of Isaiah in saying, "Lord, here am I send me." When a person asks God to use them, then they will see the process of the filling of the Holy Spirit commence. Once a believer has asked God to fill him, and has walked straight, then that believer is a candidate for the Holy Spirit to use him in the work of the Lord. Never has God filled a believer just for them to say that they had been filled with the Holy Spirit, rather, God fills believers for His purpose. God wants to fill every believer, but only a few will qualify.

CHAPTER 40
THE NUMBER THIRTY-FIVE ~ "FORSAKEN"

The Source

In Genesis 11:12, the Bible states, "And Arphaxed lived five and thirty years and he begat Salah." Salah was one of the ancestors of the Lord Jesus Christ. He is mentioned in Luke 3:35. In Genesis 11:12 the name *Salah* means "a small stream or fountain", and in the Hebrew Lexicon the word *Salah* is only found three times in the Bible. In the three times it is seen in the Scriptures, it does not have a good connotation.

The third time it is found in the Bible is in Isaiah 8:6, and it brings out the meaning of the name. Isaiah 8:6 says, "Forasmuch as this people refuseth the waters of Shiloah that go softly, and rejoice in Rezin and Remaliah's son." It says they refused the waters of Shiloah. They abandoned them for something else. His name is associated with abandonment or being forsaken.

The Symbolism

The number thirty-five in the Bible symbolizes forsaken. Jehoshaphat began his reign over Judah at the age of thirty-five years, and he reigned for twenty-five years. His mother's name was Azubah, which means forsaking. This story tells of Jehoshaphat walking with God and then forsaking God. The scriptures reveal the age of Jehoshaphat being thirty-five years when he began to reign, and it also reveals the name of his mother Azubah, which means forsaking. Jehoshaphat had a big naval adventure in which he equipped ships, linked with the wicked Ahaziah king of Israel of the northern tribe, and dispatched those ships over to Tarshish and obtained gold. God forsook him by breaking his ships, which made his navy unable to sail. The name of Seleg in Genesis 11:12 is associated with being forsaken or being abandoned. Jehoshaphat abandoned God. He walked away from God. He abandoned the way of God. The people had not prepared their hearts unto the God of their fathers, and one can tear down every idol image or false worship there is, but if the heart is not prepared for God, there will be no worship of God. A person can sit in a beautiful building, but if the heart of that person is not prepared for God, it is not good. Many assemblies have forsaken God and formed their own way of worship which is without Him. The greatest tragedy in a church today is not having a true heart for God.

The people in Jehoshaphat's day forsook God and abandoned His ways. Jehoshaphat clearly demonstrated in his life of thirty-five years that he forsook God; followed and made friendship with an evil man; and went after gold as recorded in I Kings 22:18.

The number thirty-five is not found in the New Testament. According to the book of Hebrews, Jesus said, "I will never leave you nor forsake you." Forsaken to the believer is never mentioned because in the New Testament there is a new covenant. Some people may have gotten away from God's way and God's house, but they cannot get away from God, because where they go, He is waiting there for them. He will never forsake the believer.

The Sermon

Forsaking God

I. Forsaking God Brings Correction - Judges 6:13

The children of Israel complained of God forsaking them. There are a number of verses on this. In Judge 6:13, Gideon said, "if the LORD be with us, why then is all this befallen us? and where be all his miracles which our fathers told us of, saying, Did not the LORD bring us up from Egypt? but now the LORD hath forsaken us, and delivered us into the hands of the Midianites."

The children of Israel had a history of saying, "God has forsaken us." In most cases where the Israelites thought God was forsaking them was when God allowed the enemy to enter their land and take them away captive. It was simply God bringing them under correction. It was not for punishment but for correction. God was showing them of their evil ways and bringing them back to Him. God used the Assyrians, the Babylonians, the Medes, and the Romans for this purpose. He did this in just about every generation of the Jewish people because of their sinful ways. The time of Judges was a horrible time. The Midianites had free course with Israel. They obtained anything they wanted from the children of Israel because Israel had forsaken God.

In II Chronicles 24:20, it tells that God had forsaken them because they had forsaken Him. That was in the Old Testament where an eye for an eye and tooth for a tooth was the rule and not the exception. God stated in certain terms that Israel had forsaken Him and thus He would forsake Israel.

II. The Believer Will Never Be Forsaken - Hebrews 13:5

The Holy Spirit had not been given to every believer in the Old Testament. The Holy Spirit had been given to the nation. He would come upon the prophets for them to preach and teach. They would then leave their place of preaching and return to mind the herds; work in their carpenter shop or at the pottery wheel. They would return to their work until God spoke to them again and then they would appear and proclaim the message from God. It is not like that today, for when the sinner is born again, the Holy Spirit makes His residence inside the believer. He walks with the believer now and corrects the believer instantaneously. In Psalms 9:10 the Bible says, "And they that know thy name will put their trust in thee: for thou, LORD, hast not forsaken them that seek thee". This verse is illustrated by the actions of Shadrack, Meshack and Abendigo. A great promise in the Word of God is in Psalms 37:25 where the Bible says, "I have been young, and now am old; yet have I not seen the righteous forsaken, nor his seed begging bread." The key word is "righteous." Those that are trying to live right before God, they have never been forsaken. They have always had His help.

In Hebrews 13:5 the Scriptures say,"Let your conversation be without covetousness; and be content with such things as ye have: for he hath said, I will never leave thee, nor forsake thee". Those words are not found in Matthew, Mark, Luke or John. There is a big different between *leaving you* and *forsaking you*. If a person leaves someone, they might say goodbye and state that they will see that person later. To forsake someone is to abandon them in a place of need. God said He would never abandon you in a place of need. He will never leave the believer by himself especially when they are experiencing troublesome times.

III. Some Believers Have Forsaken God - II Timothy 4:10

There have been some in the New Testament who have forsaken God and His people. In II Timothy 4:10 the Bible states, "For Demas hath forsaken me, having loved this present world.". The present world can take control of people and put a chokehold on them, as in Christ's parable of the Sower and the Seed. Some seed fell among thorns. Demas allowed the thorns and riches of this life to overcome his life. Some people backslide from the Word of God. If Demas was saved, he will be in heaven. He may not have any crowns if he has not repented and gone back to work for the Lord.

Peter tells of people in the church in his day who had gone astray. In II Peter 2:14-15 the Bible says, "Having eyes full of adultery, and that cannot cease from sin; beguiling unstable souls: an heart they have exercised with covetous practices; cursed children: Which have forsaken the right way, and are gone astray, following the way of Balaam the son of Bosor, who loved the wages of unrighteousness." One must not judge whether they were saved or not, but there is a God of grace saying that He will never leave them nor forsake them. People may be forsaken by others, but God will not forsake us.

CHAPTER 41

THE NUMBER THIRTY-SIX ~ "DEADNESS"

The Source

The number thirty-six is found only one time in the Bible in Joshua 7:5 where the men of Ai smote about thirty and six men of the Israeli army. The word "about" is in the verse because they were running from the enemy when someone was counting and they did not have time to get an accurate count. The Holy Spirit instructed the prophet to put the number thirty-six in this verse of the Holy Scriptures.

The Symbolism

The town Ai did not appear as a great threat. When the Israelis sent spies to inspect the city, they said it would not require the whole army, just give us 2,000 or 3,000 men and it would be conquered. They had based their calculations on their victory at Jericho. The word *Ai* means a *ruin or a heap.*

During World War II, the allied army bombed Germany. Stuttgart, Germany, was and is today an industrial city. The allies bombed Stuttgart heavily. After WWII, bulldozers were used to move all the demolished buildings to the center of the town. They placed dirt on the ruins of the city of Stuttgart and made a garden and a park. It is now the center of the town. Ai was a heap or ruin. When the spies looked at it, they said that nothing was there, that the whole town was a rumble or heap.

Joshua 7:5 states that the men of Ai chased the Israeli army from before the gate even unto Shebarim. *Shebarim* means a *fracture, affliction, breach, breaking, bruise, crashing, destruction and hurt.* It surely is not a good place to be. The Israeli army go from a heap; running from the enemy: and go to Shebarim. It was the acts of Achan as revealed in Joshua 7:1 that brought the defeat of the Israeli army. Achan had taken the accursed thing, and the anger of the Lord was kindled against Israel.

Someone may say that it is not fair that one man sinned and God punished the whole nation. Adam sinned and the entire world fell into sin. The people on this earth are not isolated from each other. The sins of one person or people may affect others. One member of the church commits an act and the sin brings a reproach on the whole church.

The sin of Achan was robbing God. When the Israeli army went into Jericho, Achan took the things that belonged to God. If you trace the history of the children of Israel, there were ten cities they conquered. Jericho was first and Ai was the site of the second battle. There were a total of ten places at which they fought. God distinctly said to take nothing from Jericho for it belonged to Him. Achan took what belonged to God. That was the tithe, as Achan took the first tenth of what was there. In Joshua 7:20-21, the Bible says, "And Achan answered Joshua, and said, Indeed I have sinned against the LORD God of Israel, and thus and thus have I done: When I saw among the spoils a goodly Babylonish garment, and two hundred shekels of silver, and a wedge of gold of fifty shekels weight, then I coveted them, and took them; and, behold, they are hid in the earth in the midst of my tent, and the silver under it."

Achan had no place to wear that goodly Babylonian garment. He could not wear it around the children of Israel. It looked nice. It was different than anything in his closet. He may have thought he would cut it apart and make a quilt out of it, but he was not going to wear it around town. He could not even use it. He dug a hole in his tent, buried it, covered it over thinking nobody saw it, but there were eyes in heaven that saw it. The acts of Achan brought defeat to the Israeli army. Joshua 7:24-26 states, "And Joshua, and all Israel with him, took Achan the son of Zerah, and the silver, and the garment, and the wedge of gold, and his sons, and his daughters, and his oxen, and his asses, and his sheep, and his tent, and all that he had: and they brought them unto the valley of Achor. And Joshua said, Why hast thou troubled us? the LORD shall trouble thee this day. And all Israel stoned him with stones, and burned them with fire, after they had stoned them with stones. And they raised over him a great heap of stones unto this day. So the LORD turned from the fierceness of his anger. Wherefore the name of that place was called, The valley of Achor, unto this day." Notice they raised over Achan a great heap of stones, which became his tombstone. They put up a grave marker which passed down through generations that this place was where Achan and his family are buried. It is not difficult to recognize a graveyard. You may recognize an old graveyard because of the stones or rocks used as head markers, although they may not have a name on them. Ai means a heap.

The army of Ai chased the Israelis and killed thirty-six of their army. The problem was Achan. The Israelis took him to the valley of Achor, stoned him, buried his body with all his silver and gold, and piled stones on top of him. They made a heap of Achan and his stolen belongings. The name of the town was Ai, meaning heap, and Achan is buried there today under a heap of stones. A tombstone indicates deadness in that area.

The number thirty-six is a square number. It is the product of six times six. The number six in the Bible is symbolic of man. Six times six is the sinfulness of man multiplied. The number thirty-six symbolizes deadness.

The Sermon

Deadness in a Person's Life

I. The Symptom of Deadness - Joshua 7:5

Something happened in Achan's life that caused him to become dead. He was alive to the Lord, but he was tempted and coveted a Babylonian garment. It shows how weak he was. The symptom of deadness is being dead but somehow alive. Ai was in a state of deadness. To the spies it looked like it was dead and they could conquer it easily. It was in ruins and in a heap. Joshua was convinced by the report of the spies that the city Ai could be conquered easily by an army of just two or three thousand men. The truth was that the Israeli army had deadness dwelling within their ranks because of the sinful acts of Achan.

II. The Sign of Deadness - Revelation 3:16

Lukewarmness is a sign of deadness in a believer's life. Jesus said to John that the Laodicean church was lukewarm. Jesus wished this church were either dead or alive, but they were not either. They were lukewarm. They had a deadness about them. The believer who says he is saved but lives like the world; does the things of the world; covets; goes after the Babylonian garment; the silver and the gold, then the love of Christ becomes cold in their life. People who quit loving God will quit loving His people. They will quit caring for the things of God and become dead in their walk with God.

Achan fit in real good with the Israelites around Jericho before they went into the city. Achan marched like they marched for six days and on that seventh day marched seven times. He assimilated real good, and some folks can assimilate real good in the choir, in the church, in the Sunday School class, but in their hearts they know there is a deadness about themselves. They do not have the love of God like before in their lives.

III. The Remedy for Deadness - I John 1:9

The way for the believer to overcome deadness in his personal life is to pray for cleansing for all sins. What was the downfall of Achan? Sin in his life. What caused the deadness in the Israeli army? Sin in the camp. It did not take but one man and his family to cause the people of God to lose a battle. His wife and children had to know about it. The believer needs to pray for cleansing from all sin. The believer needs to immerse himself in the Word of God. Immerse means to baptize, and to baptize is not just a little sprinkling. It means to completely be covered with water. The believer needs to submerge himself in the Word of God and listen to the Holy Spirit to be able to resist temptation.

CHAPTER 42

THE NUMBER THIRTY-SEVEN ~ "MIGHTY"

The Source

The number thirty-seven is found only one time in the Bible in II Samuel 23:39 where the thirty-seven mighty men of David are described. The number thirty-seven is a prime number, which is one of the basic building blocks of the number system. God is more aptly described and concerned in the field of multiplication than He is in the field of addition, subtraction, division or exponential function. He often uses multiplication. One times One times One is who He is. He is the Trinity. Peter told the Lord that he forgave his brother seven times when he prayed. Jesus used multiplication and told Peter that he should forgive his brother seven times seventy. The exponential function is found in the Bible in Deuteronomy 32:30, where one could chase a thousand and two could chase ten thousand. One would think that if one chased one thousand then two would chase two thousand, but not so. If two believers pray and agree together, their strength increases tremendously. When two agree as touching any one thing and ask what they will, then it will be added unto them. Where two or three are gathered together in His Name, it commands His presence.

The number thirty-seven has no twin prime number. Its twins integers or whole numbers, are thirty-five, which means forsaken, and thirty-nine which is punishment neither of which is a prime number. The number thirty-nine is not found in the Bible, but the words forty save one. The number thirty-seven stands along.

The Symbolism

The number thirty-seven symbolizes mighty. II Samuel 23:8-39 describes the thirty-seven mighty men of David. These were the men of extreme strength, courage and honor. These were exemplary men, the best in David's army. The number thirty-seven is symbolic of being mighty. They were mighty in strength and in honor.

The Sermon

Mighty Men of God

I. Mighty Works are Associated with Great Faith ~ II Samuel 23:8

As David typifies the Lord Jesus Christ, so David's mighty men typifies his followers. The first three men in this list are great examples for the church. Their lives are an example for every believer. In II Samuel 23:8 the Bible says,"These be the names of the mighty men whom David had: The Tachmonite that sat in the seat, chief among the captains; the same was

Adino the Eznite: he lift up his spear against eight hundred, whom he slew at one time." One man in a battle engaged with eight hundred men and killed them. Adino simply got his sword and killed them one by one. This man was a warrior. He was a fighter. Believers should be mighty in their works for God. Some believers are just a hit or miss. In Matthew 13:38, we are given the negative of this statement where the Bible says, "And He did not many mighty works there because of their unbelief." Jesus had faith. He could have performed any miracle, but He chose not to because of their unbelief.

There was a nobleman who came to Christ and asked him to heal his son. Jesus said that he sought after a sign. The nobleman replied that he did not. He responded that if Jesus would just speak the word his son would be healed. The nobleman knew the position of having men under his command. Christ spoke the word, and the man's son was healed and it demonstrated great faith. Mighty works cannot be done until the believer has a mighty faith in God.

II. The Believer Must Be Mighty in the Scriptures.

In II Samuel 23:9-10 the Bible says, "And after him was Eleazar the son of Dodo the Ahohite, one of the three mighty men with David, when they defied the Philistines that were there gathered together to battle, and the men of Israel were gone away: He arose, and smote the Philistines until his hand was weary, and his hand clave unto the sword: and the LORD wrought a great victory that day; and the people returned after him only to spoil." After he had killed all the Philistines; the Israelites came back to get all the spoil from the Philistine army. After the battle, his hand got weary and clave unto the sword. He had griped it so hard that he could not relax the muscle in his hand, and he could not release his hand from the sword. His hand and the sword became as one.

In Acts 18: 24 the Scriptures state, "And a certain Jew named Apollos, born at Alexandria, an eloquent man, and mighty in the scriptures, came to Ephesus." Apollos was mighty in the scriptures. Eleazar is that Christian who has clung to the Word of God, every promise, and every word, and he believed God. The enemy is in the bushes, and Eleazar is tired, but then the Philistines start coming. The children of Israel start leaving. Eleazar reaches down, grabs his sword, and starts attacking them. He does not back down. He has his sword, the Word of God, with him. He goes after the enemy and defeats the enemy. After he kills all of them and the children of Israel see that he has won, then they come back and start pilfering all the belongings of the slain soldiers, which was the warrior's right to do so. Sometimes in the mercenary army, that was the only means by which a soldier was paid. Here, Eleazar represents the believer that is strong in the Word of God. Psalms 149:6 says, "Let the high praises of God be in their mouth, and a two-edged sword in their hand."

There was a saying in World War I and World War II: Praise the Lord and pass ammunition. It says to let the high praises of God be in their voice, which is singing. The believer should carry a song for God in his heart and a two-edged sword in his hand which is the Word of God.

Hebrews 4:12 says, "For the word of God is quick, and powerful, and sharper than any two-edged sword, piercing even to the dividing asunder of soul and spirit, and of the joints and marrow, and is a discerner of the thoughts and intents of the heart." The word of God is more powerful and sharper than any two-edged sword. In Hebrews, the writer compared the Word of God to a sharp, two-edged sword.

Eleazar's hand was weak, but when he grabbed his sword and started striking the enemy to defend himself, when he finished, his muscles tightened around his sword so much that he could not let it down. He had to get his other hand and wrestled his hand loose from the sword. Eleazar could not put aside his sword because he had been using it so much. It became a part of him, like an extension of his arm. The Bible of the believer should be an extension of his life.

III. The Believer Must Be Mighty in Deeds.

In II Samuel 23:11-12 the Bible says, "And after him was Shammah the son of Agee the Hararite. And the Philistines were gathered together into a troop, where was a piece of ground full of lentils: and the people fled from the Philistines. But he stood in the midst of the ground, and defended it, and slew the Philistines: and the LORD wrought a great victory." Shammah defended his pea patch. He was not going to retreat. There was not much there, nothing but a patch full of lentils, but he stood in the midst of the ground and defended it. The Lord wrought a great victory. It was not Shammah who wrought the victory. He just defended what he thought he should do. The believer must be mighty in his deeds. Luke 24:19 says, "And he said unto them, What things? And they said unto him, Concerning Jesus of Nazareth, which was a prophet mighty in deed and word before God and all the people:" Jesus was mighty in deeds; doing the will of the Father, and living daily for God. God is not looking for someone to live for Him just on Sunday. Shammah demonstrated this characteristic in that he was mighty in his deeds by defending what was his.

CHAPTER 43

THE NUMBER THIRTY-EIGHT ~ "FEEBLENESS"

The Source

The number thirty-eight is found four times in the Bible. The number four is symbolic of man, which gives a clue to the symbolism of the number thirty-eight. It is found the first time in Deuteronomy 2:14, in which it describes the thirty-eight years that the children of Israel wandered in the wilderness from Kadeshbarnea until they passed over the brook Zered for the second time.

In I Kings 16:29, in the thirty and eighth year of the reign of Asa, King of Judah. Ahab, the son of Omri, began his reign over Israel. Ahab, the son of Omri, reigned over Israel in Samaria for twenty and two years. Ahab, the son of Omri, did evil in the sight of the LORD above all that were before him. The thirty and eighth year of the reign of Asa, King of Judah, marked the time when Ahab started his reign.

In II Kings 15:8, the Bible tells that Zachariah assumes the kingdom in the thirty and eighth year, that Azariah king of Judah did reign over Israel in Samaria. Azariah did that which was evil in the sight of the LORD, as his fathers had done. He departed not from the sins of Jeroboam the son of Nebat, who made Israel to sin.

The Symbolism

There are four occasions when the number thirty-eight is mentioned in the Bible. It is mentioned as the length of time when the children of Israel wandered in the wilderness, and God let them wander long enough until they died. The men who saw the promised land but did not have faith to enter it, wandered for forty years in the wilderness. Their trail looks like a wiggly line on a map. Time was being marked for them to die.

The Bible gives the two accounts where Ahab and Zachariah did evil. Weakness can also be a spiritual condition as it was in the beginning. In John 5:5, there is a man who has weakness in the flesh. He is feeble, and he cannot get up in time when he sees the waters troubled to arise and touch the water first. Someone else always made it to the water first because this man was so feeble and weak. The symbolism of the number thirty-eight is feebleness or being feeble. The number thirty-eight is almost the opposite of thirty-seven, which symbolizes mighty. There is a great difference between those thirty-seven mighty men of David and these feeble people associated with the number thirty-eight. They could not fight a battle. They did not have faith. They were real feeble, and feebleness is the state of deficiency of resources that requires bigger authority, force or efficiency. Feebleness can either describe a physical condition or a spiritual condition.

The Sermon
Three Feeble People

I. A Feeble People ~ Numbers 13:31

Numbers chapter 13 describes the children of Israel as they gazed into the promised land. It was a land that flowed with milk and honey, and it took two men bearing a staff to carry one cluster of grapes. The cities were walled and very great. Caleb wanted to atttack and possess it, but the men who went with Caleb reported that they were not able to go up against such strong people. The men considered themselves feeble. The people of Israel had not even seen the land or the giants, but these ten men gave such a bad report that it destroyed their faith, and they wept all night. They had seen ten plagues and ten miracles in the land of Egypt. They saw the miracle of the parting of the Red Sea, before they had arrived at the promised land. Miracles do not produce faith. A believer can see many miracles, but it will not produce faith. The Word of God says that faith comes by hearing, and hearing comes by the Word of God. It is a personal hearing of the Word of God that builds the faith of the believer. These people were feeble. They had sent twelve (the number symbolic of divine authority) to scout the promised land, but ten came back and said it could not be taken. Caleb and Joshua said that God would help to conquer this enemy. There is an interesting scripture that says God sent hornets (Exodus 23:28) into certain villages and drove out the people. God used a little army to do his work. We need not be feeble in our faith toward God.

II. A Feeble Ruler ~ I Kings 16:31

I Kings chapter 16 reveals a feeble ruler. Jezebel, the evil queen, showed more strength and power than Ahab, the King. Ahab wanted Naboth's vineyard which was next to his dwelling. He whined and cried. Jezebel questioned him, and he stated that he wanted Naboth's vineyard. Jezebel said she would get it for him, as she was a strong person. Ahab was a feeble ruler. He only had a title as king. He did evil in the sight of the Lord.

In II Kings 15, in the thirty-eighth year of Azariah king of Judah, Zachariah reigned over Israel and did that which was evil in the sight of the Lord as his fathers had done. This is another feeble man. Rulers become feeble because they do not read the Word of God and they rely on their own knowledge and their personal advisors. If the believer listens to his friends and decides to live as his friends desire, then the believer will become feeble. The believer must read the word of God and obtain convictions for himself. If the believer acts like Ahah and Zachariah, then he will become feeble.

III. A Feeble Sinner ~ St. John 5:5

In John chapter 5, the Bible reveals a man who never asked to be healed. He did the best he could for thirty-eight years. His age is never revealed but the Bible states he lived with this ailment for thirty-eight years. The grace of God is shown in this story of someone who was feeble. John 5:3 describes many believers. They are impotent, blind, halt, withered, or waiting for a miracle to change them.

In this story an angel would come down in due season and trouble the water, and whosoever was first into the water would be made whole of whatever disease he had. Evidently in the past this had happened, although this is the only account in the scriptures. God placed the number thirty-eight in this scripture for us to understand that it deals with feebleness. The idea associated with the number thirty-eight is feebleness.

When Jesus saw him lie and knew that he had been thirty-eight years in that conditions, He said unto him, "Wilt thou be made whole?" Now at other times people came to Jesus and asked to be healed, but in this instance, Jesus picked out this one person out of all the multitude. That is why it is the miracle of grace. It is a miracle of grace whenever a person hears the gospel. Titus 3:4 gives the definition of grace which is God's love and kindness toward man. This man with a physical infirmity had a feeble health record against him. The man told Jesus that he had no one to help him into the pool, and that when he came to the pool, others stepped in before him. Jesus told him to rise, take up his bed and walk, and immediately the man was made whole. He took up his bed, walked, and the same day was the Sabbath. This man was healed by Jesus by faith in the command of Christ for him to take up his bed and walk.

CHAPTER 44

THE NUMBER THIRTY-NINE ~ "PUNISHMENT"

The Source

The number thirty-nine is found two times in the Bible. The first instance is in II Chronicles 16:12 where King Asa in the thirty-ninth year of his reign developed a disease in his feet. The exact disease is not revealed but the Bible says it was exceeding great. It could have been an extreme case of gout. The tragic section of this story is that in his disease he sought not the Lord, even though he knew he was going to die. Two years later Asa died.

The number thirty-nine is also a cloaked or hidden passage in II Corinthians 11:24 where Paul is giving a history of his persecution. He relates his imprisonment. He said that of the Jews, five times he received forty stripes save one, which is thirty-nine stripes. Paul did not have a comfortable life as a preacher and carrier of the gospel, but God told him when He called him that he could suffer and endure many things. Paul was beaten by the Jewish Sanhedrin, or leaders of the synagogue, five times. The number of the grace of God is shown in these passages. Even in punishment the grace of God is present in the believer's life. Five times they beat him forty stripes save one. Under the Pharisaical law a criminal could be beaten with forty stripes. If a criminal was beaten forty-one times, then the executioner had broken the law and then he was to be beaten up to forty times as a criminal. The Jewish people would count thirty-nine strips and then stop to ensure being one short of the maximum penalty in case the count was incorrect.

The Symbolism

The number thirty-nine symbolizes punishment. The number thirty-nine is the product of three times thirteen. Three is symbolic of completeness, and thirteen is symbolic of rebellion. The number thirty-nine is the result of complete rebellion which results in punishment. When a person is in complete rebellion against God or against man, then that person must suffer the punishment.

The Sermon

Rebellion and Punishment

I. King Asa in Complete Rebellion ~ II Chronicles 16:2

In II Chronicles 16, King Asa is in complete rebellion against God and suffers the punishment. Asa was not a good king. He made alliances with the king of Syria. Asa took silver and gold out of the treasures of the house of the Lord and sent it to the king of Syria saying it was a league between them to break the king of Israel who had come against them. Taking silver and gold from the Lord's house was not acceptable. The king of Syria was an evil man who was an idolater and heathen. God sent a prophet to Asa and said that because he had relied upon the king of Syria instead of God, then the host of the king of Syria would escape from him. The prophet told Asa that if he had prayed unto God, then God would have delivered him. Asa did not rely on or trust God. Hanani preached to Asa and brought to his remembrance that he had once relied on God. He reminded him how he had defeated the Ethiopians and Lubims, and how God came to his rescue. Asa had not done that with the king of Syria. Hanani further told Asa that the eyes of the Lord run to and fro throughout the whole earth, to show himself strong in the behalf of them whose heart is perfect toward God. There is a God whom the believer can meet in prayer and ask Him to accomplish the impossible. Hanani told Asa that because he had done foolishly, from henceforth he would have wars. Asa was angry with the seer and put him in a prison house for punishment for he was in a rage against him.

Asa sowed punishment by casting Hanani into prison, and he reaped punishment when he became a prisoner himself in his own castle because of his feet. The law of sowing and reaping is no respect of persons. King Asa had some of the same beliefs as Albert Einstein who called himself a deeply religious unbeliever. Albert Einstein believed there was an order, and believed there had to be some thing or some one who created that order, but when he became pressured and asked about the Lord God Jehovah, he said that he was a deeply religious unbeliever. King Asa was the same way.

II. The Apostle Paul in Complete Rebellion ~
II Corinthians 11:24

The Apostle Paul is in complete rebellion against the Jewish leaders and suffers the punishment. In King Asa's incident, you have a man who is in complete rebellion against God. Paul was in complete rebellion against what was wrong. Because his rebellion against them, they arrested him and beat him thirty-nine times or forty save one. Notice that it does not say thirty-nine, as it is a cloaked passage. The reason thirty-nine is not given in this passage is that the believer needs to be very careful when something happens to another believer. The believer is not to be the judge of others and say that a believer is being punished by God. One person does not understand fully what God is doing in the life of another person. The believer needs to help people wherever they are in their trial, and do his best to console them. If God is chastening them, then He will do His corrective work in their lives.

Sometimes when the believer stands against evil, then the believer will suffer the consequences. Paul suffered for righteousness sake. He was punished for doing right. Five times he was whipped. No doubt the second time when his shirt was taken off his back, the punishers saw the stripes from the first beating. When Paul was beheaded on the Appian Way in Rome, an examination of his back revealed the stripes from the five beatings. He had been beaten five times thirty-nine stripes which equals two hundred ninety-five stripes. He that liveth Godly in Christ Jesus shall suffer persecution. Sometimes God allows the devil to put stripes on the backs of believers.

There are two examples in the Bible. King Asa represents rebellion against God, and he suffers. The Apostle Paul represents rebellion against the Sanhedrin, and he suffers. Just because someone suffers in life is no sign they are in trouble with God. In King Asa's incidence, it is very clear that he did not rely on God to cast out or go against the king of the north, but instead he relied on an enemy who did not care for God. When he got a disease in his feet, he did not rely on God and instead relied upon his physicians. There is nothing wrong in going to the doctor, but the believer should pray before going to the doctor. The doctor may need wisdom, and the believer should pray for the doctor to possess wisdom in curing the problem. Asa relied upon the physicians and did not call upon God in a two-year period. In seven hundred days, he could have asked God to help him with those feet, and God could have healed him immediately.

III. Punishment - Man's and God's ~ Proverbs 28:5

Men are punished for retribution or revenge for a crime. The social institutions of the world claim they put men in prison for rehabilitation to make them better. Without Christ, they are usually back in prison within a short period of time. Some men have stayed in prison so long that they know of no other lifestyle. Young men can be seen with their pants down low showing part of their underwear. This practice came from the prisons where a man would lower his pants to show his underwear to advertise his body for sale to other men in the prison. When those men were released from prison, this practice then came on the streets of our city. Most of those young men who wear their garments in that fashion probably have no idea what it represents. People are placed in prison because there is a punishment for their sin and hopefully for correction.

God punishes the believer solely for correction. God does not punish the believer to pay for his sins. A person cannot pay for their sins unless that person spends eternity in hell. Jesus paid for the sins of the believer. God punishes the believer for correction so the believer will not do the things again that will hurt or harm him. God punishes the unbeliever in hell for rebellion against him. Hell was not created for humans. It was created for the devil and his angels. When a man hears the gospel one time, the Holy Spirit deals with his heart but if he refuses to do what is right in the eyes of God, he is in rebellion against God. Christ did not come into the world to condemn the world but that the world through him might be saved.

CHAPTER 45
THE NUMBER FORTY ~ "TESTING"
The Source

The first mention of the number forty is in Genesis 7. God said it would rain upon the earth for forty days and forty nights, and every earth creature would be destroyed from the face of the earth except those on Noah's ark. The number forty is found about ninety-six times in the Bible. Beside the numbers one, three and seven, the number forty is mentioned the most. This is interesting because the number 39 is only mentioned one time in the Old Testament and in a hidden verse in the New Testament. The number 40 is a significant number because it is mentioned so many times.

Noah experienced forty days of rain (Genesis 7:4, 12). After the ark settled on Mount Ararat, Noah and his family waited forty days (Genesis 8:4-6). Jesus was tempted forty days in the wilderness (Matthew 4:2). Jesus was forty days upon the earth after His resurrection as He taught the disciples (Acts 1:2). The reigns of Saul, David, and Solomon lasted forty years. The city of Nineveh was given forty days to repent as Jonah preached (Jonah 3:4-5).

God made the children of Israel to wander in the wilderness for forty years. Everywhere they went, there were death markers or tombstones where people had died. That was the only reason God sent them there. The wages of sin is death. God could have snapped his finger and killed them at one time, but He did not. He let them examine themselves and go through this forty year period to realize the things they had done. Their evil was that they did not have faith in God. God never told them to send spies over in that land. God did not tell them to choose twelve men, send them to that land and see if it could be occupied or not. God told them to go to the promised land, conquer it, and He would be with them.

It is recorded in Judges that God sent hornets into cities, and the Israelites did not even have to fight a battle or swing a sword. They walked into a city that had been abandoned by people because the hornets had gone before them and drove out the enemy. God has a small army, but it is a mighty army. The sin of God's people was doubting God. Sometimes doubt comes by hearing from other people instead of from God. They heard the evil report from the ten men, but Joshua and Caleb said the land could be conquered. For forty years, the children of Israel wandered in the wilderness.

The Symbolism

The number forty symbolizes testing. Testing is a time of trial or a time of probation in a person's life or in a nation's existence. God gave Israel forty years to die in the wilderness. Forty is a product of five times eight, where five is symbolic of grace and eight is symbolic of a new beginning or renewal. So forty is a result of grace, along with a renewal in the believer's lives, as a result of testing during this life. God does put His people through testing. Peter likens it to the believer's life being a series of tests in which their faith is examined. In I Peter 1:7, he mentions that our faith is tried as gold put in a fire. The fiery trial purifies the gold which is faith in the believer's life.

The Sermon

The Testing of the Believer

I. The Testing of a Person's Conduct ~ Acts 7:23-30

Moses was rescued from the Nile River by Pharaoh's daughter. Miriam, his sister, was watching the baby in the basket as it floated. Moses' mother was paid to take care of her child. Moses grew and knew both the Hebrew and Egyptian language. He was educated in the Egyptian school, but he always knew he had Hebrew roots. One day he went to the Hebrew community and saw an Egyptian beating a Hebrew man, and Moses killed the Egyptian. Moses became a murderer. He tried to bury him in the sand to hide him. The next week an incident arose when one of the Hebrews asked Moses if he was going to kill him as he had killed the Egyptian man. Moses left Egypt. For forty years in Egypt, Moses was part of the royal family, but he became a murderer. He then went to Median, married, and had a family. He lived as a shepherd. After his eightieth birthday, he was in the wilderness with the sheep. He saw a bush burning and not being consumed. As he got close to that bush, a voice emanated from the bush and told him to take off his shoes for he was on holy ground. The voice from the bush told Moses that He wanted him to go back to Egypt and to lead his people out of Egypt.

Moses went to Egypt and he did lead God's people, but his life was one of trials. First, after forty years in Egypt, he committed a sin by killing a man. Second, in Median he disobeyed God not having a willing heart. The first one was a sin of commission, and the second was a sin of omission. He did not want to go, but God kept insisting. The third time he disobeyed God was at the end of this forty years when he disobeyed God by striking the rock for water. Moses was a great man, but at the end of forty years he failed God. The last mile is as important as the first. The life of Moses was broken into three forty years sections which tried his conduct.

II. The Testing of a Person's Character ~ Genesis 7:4

The testing of a person's character is the aggregate of features, traits, events and decisions that form his individual nature. Noah experienced forty days of rain. Forty days after the ark had sat down on Mount Ararat, Noah experienced the new world. When it quit raining, it had been a year and seventeen days that the waters were on the face of the earth. When the ark sat down, it was forty days later that Noah opened the window. It takes character in the believer during a time of testing. Noah and his family experienced trials that nobody else has experienced. The believer's trial may be like Noah's. If the believer has the character for forty days to listen to God and wait on Him. Sometimes waiting is a hard thing. If a believer has strong character during their trial, the believer can wait on the Lord and let God guide him. In Matthew 4, Jesus was tempted forty days by Satan. This was also a test of character. Jesus was led by the Spirit in the wilderness. It was God Who told him to go there. God wanted Him to face every temptation that a believer would ever face so that He would know how the believer felt in situations of their lives. No matter what the believer faces in life, Christ has experienced it.

The lust of the flesh, the lust of the eye, and the pride of life are the three main categories of sin, and Christ suffered those three temptations while in the wilderness. In rebuking Satan, Christ went to the book of Deuteronomy to prove His character. The believer will be stronger by relying on the Bible in a time of temptation.

In the book of Acts, the believer reads of Jesus being in His new body for forty days. This is not necessarily a testing of His character, but it shows His job and His work during that forty day period. He proved true to His character in what He was to do. He was to teach the disciples the Word of God during that time, and He started with the two men on the road to Emmaus. He taught them from Genesis to Malachi about Himself. If a believer is going through a test and trial, he is in the palm of God's hand, and the test or trial will push the believer further in His hand.

III. The Testing of a Person's Competence ~ I Samuel 17:6

In I Samuel 17:6, the warriors of Israel were fighting the Philistines and they proved competent in holding their line until God sent a deliverer, who was David. They had fought and stood the test against the Philistines for forty days. They were competent in staying on the line. Now they were afraid of Goliath, but the soldiers must receive some credit, as they did not run and hide. They did not abandon their ground. They stayed in their camp.

Sometimes the believer must stay his ground and wait on God to move in their situation. David, the shepherd boy, heard the giant cussing, screaming, and calling God all kinds of names, and David said, "Is there not a cause?" His brothers wanted to send him home, saying he was just a bread and cheese carrier. Some said Goliath was so big that he scared the other men, but to David he was only a big target. David cast a rock, and the Holy Ghost guided it right to the middle of his head. It sunk into his forehead, and he fell forward. Every knee shall bow and every tongue shall confess. The soldiers of the nation of Israel had stood against the enemy for forty days. Sometimes the believer has to stand where they are, until God sends somebody to help win the war in their cause. The soldiers in Saul's army stood competent against the line that the Philistine warriors had presented for forty days.

The reigns of David and Solomon lasted forty years. Their competence or incompetence was proven during this time. II Samuel 5:4 describes David's reign on the throne for forty years. For a king to stay on the throne ten years was a difficult task, for the king had made enemies. The enemies would do everything they could to remove the king from the throne. Saul stayed on the throne for forty years. He proved his incompetency toward the end of his reign. David and Solomon both ruled for forty years and served their purpose. The believer should stay where they are until God is through with His work.

In the book of Jonah, the city of Nineveh was given forty days to repent. They proved competent in making the right decision. When they repented, Jonah lamented that he had gone there, because he said that he knew God would forgive them. He was mad at God for giving a revival. The forty day period in Nineveh displayed Jonah's heart.

CHAPTER 46
THE NUMBER FORTY-ONE ~ "LEADERSHIP"

Source

The number forty-one is found only five times in our Bible. The first mention is in I Kings 14:21 where the Bible says, "And Rehoboam, the son of Solomon, reigned in Judah. Rehoboam was forty and one years old when he began to reign, and he reigned seventeen years in Jerusalem." The number forty-one is found also in I Kings 15:10, II Kings 14:23, II Chronicles 12:13, and II Chronicles 16:13. Five times in the Bible this number is found, and these verses describe three rulers in the Holy Land.

Symbolism

The number forty-one symbolizes leadership. These were three leaders. Two were leaders of the southern kingdom, and one was leader of the northern kingdom. Leadership is the ability to lead, give guidance and direction, whether good or bad.

The number forty-one is a prime number. It has a twin prime, which is the number forty-three. Leadership is a key component in a church but no man is a leader walking by himself. A leader leads people. Leadership is very important in leading a nation and leading a church. There are three men in the Bible who are associated with the number forty-one. Each man was a ruler in the Holy Land.

Sermon

Leadership

I. A Waning Leadership ~ II Chronicles 12:1

Rehoboam displays a waning leadership. His rule was good in the beginning, but after three years he wavered. II Chronicles 12 says that king Rehoboam strengthen himself and began his reign in Jerusalem when he was forty-one years old, and he reigned seventeen years. The number seventeen in the Bible is symbolic of victory. Rehoboam did evil because he prepared not his heart to seek the Lord. He was king over the southern kingdom, and the son of Solomon. For three years Rehoboam lived for God and walked in the ways of David and Solomon until his kingdom was strengthened. In the fifth year of his reign, Shishak king of Egypt came up against Jerusalem and took away the treasures of the house of the LORD, and the treasures of the king's house. He took away all the shields of gold which Solomon had made. King Rehoboam made brazen shields to replace the golden ones, and committed them unto the hands of the chief of the guard, which kept the door of the king's house. Rehoboam's kingdom went from gold to brass. Gold is the symbol of faith which is believing God and trusting

His Word. Brass is symbolic of judgment, and Rehoboam was judged by God. He had twelve more years during his reign under the judgment hand of God. God in his mercy allowed him to live twelve more years, but Rehoboam had a waning leadership, or a leadership without God in control. There are two kinds of leaders in the world, those who are interested in the fleece, and those who are interested in the flock.

II. A Weakened Leadership ~ I Kings 15:10

In I Kings 15:10, King Asa reigned for forty-one years. He displayed leadership by staying in power for a lengthy time, but it was a weakened leadership. There were others who wanted to be king, and sought ways to get him out of the way, but King Asa survived. Asa did that which was right in the eyes of the LORD, as did David his father. He took away the sodomites out of the land, and removed all the idols that his fathers had made. He removed his mother, Maachah, from the throne, because she had made an idol in a grove. King Asa destroyed her idol, and burnt it by the brook Kidron. Even though the high places were not removed: nevertheless Asa's heart was perfect with the LORD all his days. King Asa was the grandson of Rehoboam and seemed to be a good man. In II Chronicles 16, the Bible tells that he did not rely on God like he did in his first years. King Asa had a disease in his feet and sought not the Lord, and he died in the forty-first year of his reign. This showed a weakened leadership. It went from faith to sin. There is more to sin than just committing some horrible act. James 4:17 says, "To him that knoweth to do good and doeth it not, to him it is sin." There are sins of commission and sins of omission. The sin of King Asa was in his battle against the northern kingdom. He went to the king of Syria and asked for his help and did not pray unto the Lord for His help. Yet a few years before that time, in his battle with the Ethiopians who had an army of one hundred thousand men, Asa called upon God and God turned them back. Yet when it came to fighting the northern kingdom, he did not call upon God. He became weak in his leadership. He went from faith to sin, and when Hanania the seer came and told him that he had done wrong, King Asa put him in prison. It was just a year or two later that he developed a disease in his feet, and after two years he died because he never called upon God to heal or help him. He had a more horrible condition than just his feet; he became sinful. A weakened leadership is one that does not rely upon God. The world would be happier if its leaders had more dreams and fewer nightmares.

III. A Winning Leadership ~ II Kings 14:23

In II Kings 14, Jeroboam began to reign in Samaria, and reigned forty-one years. He did not live for God, but strangely enough God used his kingdom to restore to the nation of Israel valuable land and people. God called Jonah, Hosea, Amos and Micah to prophesy during the reign of Jeroboam in the northern kingdom. Jeroboam was not a good king. Even though he did evil, God still used him.

The people of God were overrun by the Syrians. They were being persecuted, and it troubled God's heart. God used a wicked leader to recapture the land and free the children of Israel who were in a bitter affliction. God can use strange leadership to accomplish His purpose. A real leader faces the music even when he dislikes the tune.

The number forty-one deals with three men totally different in their leadership style. One starts out strong, but he goes from gold to brass. One starts out with a strong leadership, but then he becomes weak along the way because he does not have enough faith in God. He goes from faith to sin. One leader brought victory to the land of Israel, even though it did not appear he lived for God.

CHAPTER 47

THE NUMBER FORTY-TWO ~ "CONDEMNATION"

The first mention of the number forty-two is in Numbers 35:6 here the Levites were instructed to take possession of six cities for refuge, and they were also given forty-two other cities. II Kings 2:24 is an interesting story about Elijah, who had been carried by God with chariots of fire into heaven. Elijah was separated from Elisha. Elisha then traveled to the school of the prophets, and on his way he went through Bethel. When the prophet approached Bethel, children came out of the city and mocked him. They said, "Go up, thou bald head. Go up, thou bald head." He turned back and looked on them and cursed them in the name of the Lord, and there came two she bears out of the woods. The bear tore forty and two children. Elisha then went to Mt. Carmel and then turned to Samaria.

In this story, there are forty-two children who mocked the man of God. They had probably heard of the catching away of Elijah, and mocked Elisha in prodding him to go up also. They said, "Go up, thou bald head." It was a mockery. They were making fun of him because he had followed Elijah and now he was alone. The school of the prophets had gone to the woods to look for the body of Elijah thinking that he had fallen back to earth. The children heard about the story of

Elijah's catching away from their parents and people of the city. Two bears came out of the woods and killed forty-two children of the city. Condemnation in is this story. In II Kings 10:14, Jehu and Abner are rivalry generals from the northern and southern kingdoms. Jehu met with the brethren of Ahaziah, King of Judah, and said, Who are ye? They answered, "We are the brethren of Ahaziah; and we go down to salute the children of the king and the children of the queen". Jehu said, "Take them alive". And they took them alive, and slew them at the pit of the shearing house, even two and forty men; neither left he any of them. He killed the last remaining forty-two relatives of King Ahab. In Revelations 11:2, the gentiles shall tread under foot Jerusalem for forty-two months. God could have said about 1,650 days or He could have used a smaller number, but he used forty-two months. In Revelations 13:5, power is given to the beast for forty-two months. God could have put that in number of days, and it would have been a totally different number, but He chose the number forty-two.

The Symbolism

The number forty-two symbolizes condemnation. In Numbers 35:6, there were six cities of refuge. One could not hide in the other forty-two cities and be safe. There were three cities on one side of the river and three cities on the other side of the river so that crossing the river was no excuse. If a person had killed someone accidentally, and went to a city other than one of those six cities, the person could not proclaim yourself as innocent before the priest and get a reprieve. He could not get a new name, a new white stone, and a new residency. The cities of Refuge were God's protective program during that time for people who had killed someone innocently. It was a time of an eye for an eye and a tooth for a tooth, whether it was innocent or not. If a person went to one of these other forty-two cities and tried to hide there, it was in vain because it was a non-refuge city. For the slayer it was a city of condemnation.

In II Kings 2:24, forty-two young people were killed by two bears for mocking Elisha. These young people were under the condemnation of God for mocking the man of God.

In II Kings 10:14, there is a man thinking he is doing the will of God when he lures in King Ahab's last remaining male relatives and he kills them. He is under the condemnation of the ruler for disobeying his wishes.

In Revelations 11:2, the Gentiles will be in control of Jerusalem for forty-two months, and there will be condemnation in Jerusalem for forty-two months. In Revelations 13:5, power will be given to the beast for forty-two months, and he will execute his enemies like Hitler and Stalin. Here is clearly a picture of condemnation.

Mathematically the number forty-two is a product of two and twenty-one. The number two is symbolic of witness and the number twenty-one is symbolic of evil. The witness given to the evil is condemnation, a soon coming endurance of something unpleasant or undesirable.

The Sermon

Life and Death

I. The Place of Life - Nehemiah 35:6

In Nehemiah 35:6, there is a place of life. If a killer ran to a city of refuge (Numbers 35:6), he was to present himself to the priest. The priest then questioned the person to determined his guilt or innocence in killing another person. If the priest determined the person was killed under premeditation, then he would turn the man over to the relatives of the murdered man for punishment. It was not an automatic refuge. If a man had committed the act innocently, then that man was to run as fast as possible to one of these cities of refuge. Then the priest under the Israeli protection program, would take that man to a house, change his name, give him a white stone with a new name on it to put in front of his house as a marker. When the brothers of the dead man came to find the killer, they could not locate him because he had a new identity. If the killer had gone to the other forty-two cities and went in there, the priest had no power to protect him. He would be under condemnation.

II. Death Has a Stronghold on Mankind ~ Ezra 2:24

In Ezra 2:24, the name of Azmaveth's means "a strong one of death." Condemnation brings forth death. Death has a stronghold on mankind. There have only been two people in the Old Testament who did not see death. Even the great leader Moses saw death. Enoch did not see death, for God translated him out of this world. Elijah did not see death, for God called him also out of this world. For everybody else except one person, death still has a hold. That one person is our Lord Jesus Christ who defeated death. On the third day, Jesus arose victorious out of the grave. Death did not have that stronghold on Jesus. He was a man but he did not have the sin-tainted blood of a man. He had God's blood.

III. The Wages of Sin is Death ~ II Kings 10:1

In II Kings 10, when the letter came to the rulers of Israel; they took Ahab's seventy sons, and slew them. They put their heads in baskets, and sent him them to Jezreel. And there came a messenger, and told him, saying, "They have brought the heads of the king's sons." And he said, "Lay ye them in two heaps at the entering in of the gate until the morning." And it came to pass in the morning, that he went out, and stood, and said to all the people, "Ye be righteous: behold, I conspired against my master, and slew him: but who slew all these? Know now that there shall fall unto the earth nothing of the word of the LORD, which the LORD spake concerning the house of Ahab: for the LORD hath done that which he spake by his servant Elijah". Through Elijah, God had placed a condemnation upon the house of Ahab. Elijah had said that Ahab and his generation would all die and be purged from the earth." The descendants of Ahab were under condemnation and it was sin that killed the descendants of Ahah.

CHAPTER 48

THE NUMBER FORTY-THREE ~ "COUNSEL"

The Source

The number forty-three is found indirectly in II Chronicles 22:2. Ahaziah was forty-two years old when he began his reign, and he reigned only one year in Jerusalem. This man was forty-three years old when he passed away. He received counsel, although it was not good counsel. The predominant theme of these verses is the word *counsel*. Ahaziah received council. It was not good counsel, but rather evil and wicked counsel. A person in a leadership position needs good counsel about future plans. Many times good persons make bad decisions and become disruptive because they have evil or bad counsel. The number forty-three is also found in II Chronicles 22 where King Ahaziah's mother was his counselor. His mother's name was Athaliah the daughter of Omri. Omri also walked in the ways of the house of Ahab for his mother was his counselor to do wickedly. He had one of the greatest grandfathers in the Bible, Jehoshaphat. Because he was the son of Jehoshaphat, who sought the LORD with all his heart, when they slew him they buried him. His grandfather was such a good man, that when this wicked man was killed in battle, they wanted to bury him. In those days the enemy would leave the body on the battlefield, and let the vultures pick the meat off the body, but because of Omri's grandfather, they buried him. Ahaziah got bad counsel from his mother. The same counselors which advised King Ahab were now Ahaziah's counselors. He did evil in the sight of the Lord like the house of Ahab had done. His mother gave him bad advice. All the secretaries of state, commerce and other cabinet positions who served Ahab and directed him in ungodly ways were now the advisors of Ahaziah. Ahaziah took all the wrong steps and did everything wrong. Three times the word counsel or a form of the word counsel is used to describe Ahaziah's life and how he died.

The Symbolism

In searching for the idea associated with the number forty-three in the Bible, the believer will find three different times the word counsel or counselor is used, and the search for the symbolism of the number forty-three reveals *counsel.* Leadership and the right counsel produce a good leader, and this was certainly not the case in Ahaziah's life. Counsel is the advice given especially as a result of consultation. The believer should seek God for counsel. The answer of God will be a peace given by the Holy Spirit. He can give the believer peace of heart, peace of mind, and a peace within the soul about a subject. God's advice is the best advice the believer should seek, not the advice of another human. When the believer receives the peace of God about a situation, then God is going to bless that believer.

The Sermon

Two Kinds of Counsel

I. There are Two Kinds of Counsel ~ Exodus 18:19

There is a Godly counsel, and there is an ungodly counsel. Godly counsel is found in Exodus 18:19, where it says, "Hearken now unto my voice, I will give thee counsel, and God shall be with thee: Be thou for the people to God-ward, that thou mayest bring the causes unto God." There is a Godly counsel that God wants every believer to hear. The believer needs to know the Bible. The believer needs to walk according to the Bible so that he will be an example of Godly counsel. When a person reads the Bible, that person is hearing God speak to him, and then he has to decide if he is going to choose to walk that way or not. When the believer walks the way of the Bible, he are taking Godly advice. In Psalms 33:11 the Bible says, "The counsel of the Lord standeth forever, and the thoughts of his heart to all generations." When a man takes God's counsel, it will stand forever.

There is ungodly counsel as shown in Psalms 1:1 where the Bible says, "Blessed is the man who walketh not in the counsel of the ungodly, nor standeth in the way of sinners, nor sitteth in the seat of the scornful." There is an ungodly counsel that people can walk in. The worst thing the believer can ever do in a spiritual warfare is go to the world for counsel. If someone is upset with their spouse and thinking of divorce, instead of talking to the minister or arranging a time with a Christian counseling center, that person will go to work and talk to somebody who has been divorced several times seeking counsel. That person has received bad advice from the world. The believer must be careful about going to the world for advice. Godly counsel is worth more money than you could ever pay to receive it. Ungodly counsel is free. The believer needs to hear Godly counsel so he can be pleasing to God, and he needs to refuse ungodly counsel. Blessed is the man that walketh not in the counsel of the ungodly. The believer should not listen to the ungodly for their advice. Advice is like mushrooms. The wrong kind can prove fatal.

II. The Fruit of Good Counsel ~ Proverbs 11:14

In Proverbs 11:14 the Scriptures says, "Where no counsel is, the people fall, but in the multitude of counselors there is safety." Good counsel equals safety. The believer will make the right decisions and go the right direction if he listens to good counsel. The fruit of counsel is safety in your life. Good counsel brings wisdom. Psalms 19:20 demonstrates that we are to hear counsel and receive instruction so that we may be wise in that latter end. Being wise is the fruit of taking good counsel. When the believer takes the counsel, and applies it to his life, and then the believer will discover that he has gained wisdom. Without counsel, purposes are disappointed, but in the multitude of counselors they are established.

III. The Need of Good Counseling ~ II Chronicles 22:4

The account of King Ahaziah is given for every believer's admonition. The Old Testament stories are given as an example to the believer whether as a good precedent or a bad one. Every believer needs good counseling. Advice is like medicine - the correct dosage works wonders, but an overdose can be dangerous. The believer can get counsel from many good sources in this life, but he should first seek God. David said that the believer needs to hide God's Words in his heart so he might not sin against God. The Holy Spirit will bring forth verses of the Bible when the believer needs them. The Bible and the Holy Spirit provide the best counseling the believer will ever receive.

CHAPTER 49
THE NUMBER FORTY-FOUR ~ "HELL"

The Source

The number forty-four is not found in the Bible, but the symbolism can be derived from its factors and location in the list of numbers and their symbolism. Forty-four has factors of 1, 2, 11, 22 and itself, 44. These are relevant to determining the symbolism of the number forty-four.

The number forty-four has a twin number of forty-two, although neither are prime numbers. The twin of a number is a number separated by just one other number.

The Symbolism

The number forty-four symbolizes Hell. The number forty-four has factors of 1, 2, 11, 22, and 44. The symbolism of two is witness. The symbolism of eleven is judgment. The symbolism of twenty-two is light. When a person receives a witness of the ways of God, then there is light or an enlightenment given to that person. If a person receives a witness (2), and a witness (2) again, then there is judgment (11). When someone has received a witness of the love of God and refuses that witness, then there is a judgment and then hell is the destiny of that person. We know that two times two times eleven is forty-four. God seems to utilize multiplication rather than the other three functions of the number system. One plus one plus one equals three, but one times one times one equals one which is a picture of the trinity of God. When Peter bragged that he had forgiven his brother seven times in one day, Jesus used multiplication when he stated that Peter should forgive him seven times seventy (Matthew 18:22). God used multiplication to illustrate the principle of forgiveness.

The twin number of forty-four is the number forty-two which symbolizes condemnation. These two numbers, forty-two and forty-four, walk together just as condemnation and Hell walk together. Hell or eternal punishment is called the second death which is enduring, something unpleasant and undesirable (Revelation 20:14).

The question of why the number forty-four is not found in the Bible may puzzle the Bible reader, but the things God leaves out of His Word are just as important as the things He places in His Word. In the six days of creation in the book of Genesis, God saw that "It was good" in regards to every day of creation, except the second day. He simply says, "It is so" (Genesis 1:7) concerning this particular day of creation. He did not say, "It was good" over this day as He did the other five days of creation. The author contends that this was the day that God prepared Hell for the devil and his angels. After preparing

Hell, how could God say, "It was good?" Even though God had just finished creating the wondrous heavens, He did not say, "It was good." The things God has left out of the Bible are just as important as the things He has placed in His Word.

Of all the animals in the Bible the cat is not mentioned. There is no mention of God in the book of Esther. The believer does not know specifically the physical problem the Apostle Paul had in his body. Many things are not mentioned in the Bible but they are missing for a divine reason.

The Sermon

One In Hell, Five on the Way

I. No Hope in Hell ~ Luke 16:24

The rich man in hell asked for a drop of water. He did not ask to be released from this place of torment because there was no hope of his escape. Even in Hell, the rich man still wanted to command people as he had on earth. He told Abraham to command Lazarus to bring a drop of water and place it on his tongue. His call for a drop of water illustrated there was very little change of his nature.

II. No Help in Heaven ~ Luke 16:27-31

One was in Hell and five more were on the way. Since Abraham would not give permission for Lazarus to visit the rich man and place a drop of water on his tongue, the rich man requested that Lazarus would visit his father's house and warn his five brothers lest they would come to the place of torment. Abraham gives the rich man some hard news. God would not send someone from the dead like Lazarus to warn them. The rich man in hell desired God to produce a miracle so that his brothers would escape the place called Hell. The rich man did not even understand his brothers much less all of mankind. God would not send someone from the dead because they would not listen to them.

III. No Hope Without Hearing ~ Luke 16:29

Abraham proclaims to the rich man in Hell that his five brothers have Moses and the Prophets to hear. Abraham tells the rich man in Hell that his five brothers must hear the preachers of their day. There is no hope without hearing the gospel. Unless someone hears the gospel, believes it and obeys it, then there is no hope. One was in Hell and five were on the way. Hell will be populated by two classes of people, those who will do anything and those who do nothing.

CHAPTER 50
THE NUMBER FORTY-FIVE ~ "PRESERVATION"

The Source

In Genesis 18:28, Abraham questions God the second time if He will save Sodom if He finds forty-five righteous persons there. Caleb testifies that he has waited forty-five years to go into the promised and take it for a possession in Joshua 14:10. Solomon's summer home according to I Kings 7:3, had forty-five pillars of cedar which were configured in three rows of fifteen pillars each.

The Symbolism

The number forty-five symbolizes preservation. The city of Sodom would have been spared or preserved by God if forty-five righteous persons were found in the city. The city of Sodom could have been preserved because of forty-five men. Caleb stated that God had preserved his life and his strength for forty-five years to conquer the enemy. Cedar which was used in Solomon's summer house is one of the most desirable woods when preservation is desired. Cedar beams would last longer than any other common wood. Forty-five is the product of five times nine. Five is symbolic of grace and nine is symbolic of divine completeness which equals preservation. Three is symbolic of complete and fifteen is symbolic of rest.

The Sermon

The Force of Preservation

I. The Preserving Force ~ Matthew 5:13

The righteous believer is the preserving force in this world. The early believer understood that salt was used to preserve almost all of their meats and many vegetables. Olives placed in a brine solution was a common practice during the times of Jesus. Paul told the believers at Colosse to have their speech always with grace, seasoned with salt, so they would know how to answer every man. The believer is a preserving force.

II. The Purposeful Force ~ Joshua 14:10

Caleb endured the death march through the wilderness because he was preserved for a purpose. He was to be a leading force in conquering the Promised Land. God could have used many others in the place of Caleb, but God preserved Caleb to do a work for Him because of his faith.

III. The Peaceful Force ~ I Kings 7:3

Solomon's summer house was a place of peace. Solomon retreated there during times to secure a rest in his life. The cedar pillars were reminiscent of the peace that Solomon had accomplished during his kingdom. Like the forty-five cedar pillars, Solomon was a peaceful force in the nation of Israel during his reign. The believer can also be this force of peace.

CHAPTER 51
THE NUMBER FORTY-SIX ~ "LEGACY"

The Source

The number forty-six occurs once in the Bible as revealed in St. John 2:20. The Jews had questioned Jesus, hoping to entrap Him, by asking, "What sign shewest thou unto us, seeing that thou doest these things?" The Jews did seek a sign which was common to their religious thinking. When Jesus responded, His answer was completely misunderstood. He stated, "Destroy this temple and in three days I will raise it up." He, of course, was speaking of His body, the temple of God. They thought that He was speaking of the temple built by Herod. Then said the Jews, "Forty and six years was this temple in building, and wilt thou rear it up in three days?"

The Symbolism

The number forty-six accounts for the number of years it took Herod to build the temple of God. Herod took pride in the buildings he engineered and was a builder of magnificent buildings in his days. Ironically, Herod was one of the most despised rulers over the nation of Israel, yet he did a work for them by building one of the most beautiful buildings, the temple of God. Forty-six is symbolic of a work of legacy.

Forty-six is the product of two, which is symbolic of witness and twenty-three is symbolic of death. After death, the witness of the believer as a Christian will be determined to be a legacy or not. The believer must stand before the Judgment Seat of Christ.

The Sermon
God's Plan for Your Life - A Work For Him

I. God Has a Desire for Your Life - II Timothy 2:20-21

God's desire for every believer is that they would be a vessel of honor. Every parent aspires for their children to be respectable. God desires the same for His children.

II. God Has a Design for Your Life - Philippians 1:6

God has a design and plan for the believer's life. Paul was confident that God who had started a work in the lives of the believers at Philippi would also perform or complete it. Through prayer and the leading of the Holy Spirit, the believer can follow the path God has designed.

III. God Has a Destiny for Your Life - I Corinthians 3:10-15

God's end result for the believer's life is to glorify Him by the life the believer has lived. Will the believer's life be one of gold, silver and/or precious stones? That is the destiny that God desires from every child of His.

CHAPTER 52
THE NUMBER FORTY-SEVEN ~ "DISAPPOINTMENT"

The Source

The number forty-seven occurs indirectly in the Bible in II Chronicles 24:1 where Joash was seven years old when he began to reign and reigned for forty years. Therefore, Joash was forty-seven when his reign ended. This was the length of his life. He was a descendant of King David. The story of King Joash is interesting. He was preserved as a child in II Chronicles 22:10-12. At seven he was crowned as king in II Chronicles 23:11. He repaired the Temple which was a blessing to the children of Israel (II Chronicles 24:4). In II Chronicles 24:15, the high priest, Johoiada, died. In II Chronicles 24:17-18, King Joash forsook God. To redeem King Joash back to Him, God send the son of the high priest, Zechariah to restore him. In retaliation, King Joash ordered Zechariah to be killed in II Chronicles 24:22 which was mentioned by Jesus in Matthew 23:35.

The Symbolism

The number forty-seven in the Bible symbolizes disappointment. It truly describes the condition of King Joash at the end of his forty-seven years of life. He had been preserved by God, but proved to be a disappointment to God.

The Sermon

Don't Disappoint God

I. The Fight is Immutable - Job 2:3

The fight against evil is immutable; it never changes. Job was described by God as a man who escheweth evil. He continually abstained from evil. The fight against evil never changes for the believer. He must constantly fight.

II. The Failure is Impermanent - Proverbs 24:16

Solomon stated that a just man, even though he may fall seven times, he rises up. Failure for the believer may be an expensive learning experience, but he will rise to fight against evil again. In the fight against evil, the believer can arise.

III. The Finish is Important - II Timothy 4:7

The start for the believer is important, and the finish is important. The believer is in a fight against evil and the fight requires a good finish. The last mile is just as important as the first mile in the believer's life.

CHAPTER 53
THE NUMBER FORTY-EIGHT ~ "SACRED"

The Source

Two times in the Bible, the number forty-eight occurs. It is found in Numbers 35:7 where the Levites are given forty-eight cities in which to dwell. In Joshua 21:41 the promise given in Numbers 35:7 is fulfilled with the Levites occupying the forty-eight cities. The two times the number is mentioned in Scripture is dealing with the sacred cities in which the holy men of the tribe of Levi dwelt.

The Symbolism

The number forty-eight describes the number of cities which were given to the Levites. The Levites were the descendants of Aaron, the brother of Moses. They were given the task of interceding before God in a holy manner. In these forty-eight cities, the priests prayed unto God and studied His Word. This devotion was found more in these forty-eight cities than in the other cities of the children of Israel. The idea associated with the number forty-eight is sacred. The Levites were not given a specific area of land to inherit, rather they inherited these forty-eight cities. Six of these cities were designated as cities of refuges (Numbers 35:6). The forty-eight cities were sacred to God.

The Sermon
Be Ye Holy

I. Sacred in Body (Sanctified) - I Corinthians 6:19

The believer needs to realize that his body is the temple of the Holy Ghost which is in the believer. The body of the believer is a holy place. It is sacred. It is sanctified or set apart for the use of God. Be ye holy!

II. Sacred in Building (Sanctuary) - Ezekiel 44:23

There are buildings and plots of earth which have been designated as sanctuaries. The sanctuary or place of worship for the believer should be considered holy. Ezekiel tells the difference will be taught between the holy and the profane. The difference between the clean and the unclean will be taught. The believer needs to realize that some places are holy and some places are unholy. Be ye holy!

III. Sacred in Beliefs (Sinless) - Matthew 15:18

The believer needs to realize that the words and expressions which come out of his mouth are the things which can defile him. Christ clearly stated that it was not the things which went into a man that defiled man, but the things which came out of the mouth defiled the believer. The believer needs to be sacred in his beliefs and shun evil. The believer is to be sinless in his actions and words. Be ye holy!

CHAPTER 54

THE NUMBER FORTY-NINE ~ "FAITH"

The Source

The number forty-nine is found one time in the Bible in Leviticus 25:8. Leviticus 25 describes the requirements for the year of Jubilee. Every seventh year was to be a year of freedom for the children of Israel. The seventh year was a sabbatical year in which the land was allowed to rest. No grain or fruit was to be harvested during the seventh year, but God promised that there would be meat to consume (Leviticus 25:7). The children of Israel were told to number the years, seven times seven years. The forty-ninth was a year of freedom, and the fiftieth year was the year of Jubilee. There were two consecutive years when the children of Israel did not plant or harvest, but relied on the promise of God that He would provide meat and food for them to eat.

The number forty-nine is the product of seven times seven. Seven is symbolic of perfection. The number forty-nine is the product of perfection times perfection.

The Symbolism

The number forty-nine describes the periods of yearly sabbaths that God requested His people to observe. In their lifetime, the average Israeli would observe one or two Jubilees.

Forty-nine symbolizes faith. The children of Israel had faith that God would take care of them during these years. The teenage Israeli would perhaps question his parents why they were not planting crops that year. The parent would tell their child that God would provide what they needed that year. The faith of the Israeli child would be increased just by observing God as He worked in the year of Jubilee and the years before the Jubilee.

Observing the Sabbath periods brought faith to the children of God just as giving the tithe was an act of faith that was rewarded. In Leviticus 25:20-22, God gives His Word to the children of Israel that He would increase the blessings upon them in the sixth year, and it would bring forth fruit for three years. The idea associated with the number forty-nine in the Holy Scriptures is one of faith in God. Faith in God is rare in the Christian realm, which might explain why the number is found only one time in the Scriptures.

The Sermon

Faith

I. The Act of Faith - Leviticus 25:1-8

The act of faith is not some blind deed. Faith is not blind. It is hearing God's Word, and following it. God commanded the children of Israel to work for six years, and then to allow the land to rest for one year. God promised to provide food.

II. The Nature of Faith - Job 23:10

Job knew that serving God was right, and he was going to take that path, irregardless of what others said. Job walked according to God's Word which was an act of faith. The nature of faith is like a precious metal being placed in a fire. Peter stated that the trial of the believer's faith is much more precious than gold which has been purified by fire (I Peter 1:7). Gold is symbolic of faith in the Bible. The number forty-nine is symbolic of faith in the Scriptures. The faith of the believer is tried like gold placed in a fire.

III. The Reward of Faith - Leviticus 25:21

By walking by faith according to the Word of God, God promised the children of Israel that He would command His blessings upon them. He promised that the earth would bring forth fruit in one year to provide them food for three years. The reward of faith is more precious than of gold. At the judgment seat of Christ, the works of every believer will be tried by fire. The first thing mentioned is gold. Faith is one of the elements which God is looking for in the lives of the believers. God will reward the believers according to their faith.

CHAPTER 55
THE NUMBER FIFTY ~ "HOLY SPIRIT"

The Source

The number fifty is found in Genesis 18:24, where Abraham, in his conversation with God, searches for fifty righteous men in Sodom and Gomorrah. The curtain of the Tabernacle had fifty loops in the top to suspend it (Exodus 26:10-11). Leviticus 23:16 describes the Feast of Pentecost, while Leviticus 25:10 - 11 explains the Year of Jubilee which occurred every fifty years..

The Symbolism

The idea associated with the number fifty in the Scriptures is the Holy Spirit. The upper limit of fifty righteous men was the highest number with which Abraham started his argument with God concerning the destruction of Sodom (Genesis 18:27-33). The fifty loops in the linen curtain surrounding the Tabernacle symbolized the work of the Holy Spirit in the work of Christ on the cross. The manufacture of linen was a beating process, and so was the work of Christ on the cross. In Leviticus 23:16, there were fifty days after the sheaf offering till the Day of Pentecost. Fifty days after Christ arose from the grave, the Holy Spirit filled the Upper Room and the disciples on the Day of Pentecost (Leviticus 23:16; Acts 2:1). Great freedoms were granted during the Year of Jubilee (Leviticus 25:11-17). Freedom is given by the Holy Spirit at the moment of salvation.

The Sermon
The Work of the Holy Spirit

I. The Holy Spirit Produces Righteousness in the Life of the Believer ~ Genesis 18:24

The Holy Spirit reproves sin in the life of the believer. The Holy Spirit guides the believer in the path of righteousness. Abraham's argument was if there were fifty righteous men living in Sodom, God should spare the entire city. Genesis 18:26 reveals that God agreed with this proposition. The righteousness which the Holy Spirit produces in the life of the believer is a stabilizing effect in the community where the believer lives.

II. The Holy Spirit Connects the Work of the Cross and the Grace of God ~ Exodus 26:10

The Holy Spirit connects the work of Christ on the cross to the grace of God. The fifty loops on top of the Tabernacle typify the work of the Holy Spirit. The linen curtains typify the grace of God. Linen was beaten to be pliable to be used to weave the curtains. The body of Christ was beaten for the iniquities of the believer. The loops in the curtains held them.

There were two groups of five linen curtains, and the number five symbolizes the grace of God.

III. The Holy Spirit Gives the Believer Freedom From Sin ~ John 8:36

Just as the Year of Jubilee gave great freedom to the child of Israel, the Holy Spirit through the work of Christ on the cross makes the believer free. The believer is free indeed.

CHAPTER 56

THE NUMBER SIXTY ~ "STRONG"

The Source

The first mention of the number sixty in the Scriptures is found in Genesis 25:26 when Isaac was sixty years of age. This was the time when his twin sons, Jacob and Esau, were born. In Deuteronomy 3:4, the Israeli army captured sixty cities of the enemy. Solomon ruled sixty great cities in his kingdom (I Kings 4:13). The daily provision of meal for Solomon and his court was sixty measures of meal (I Kings 4:22). The Temple which Solomon built for God was sixty cubits in length (I Kings 6:2). Sixty Israeli men were taken to the King of Babylon and killed (II Kings 25:18-21). Solomon had sixty guards around his bedchamber (Song of Solomon 3:7). The distance from Jerusalem to Emmaeus is sixty furlongs (Luke 24:13).

The Symbolism

Sixty in the Scriptures symbolizes being strong or having strength. God told Isaac that two children were in Rebecca's womb and that one would be stronger than the other (Genesis 25:23). The cities captured by Israel in Deuteronomy 3:4-5 were strong with high fences, gates, and bars. The sixty cities ruled by Solomon were strong cities with walls and brasen bars

(I Kings 4:13). The sixty measures of meal eaten by Solomon and his court gave them strength (I Kings 4:22). The King of Babylon killed the strongest men of the city (II Kings 25:18-21). The strongest men guarded Solomon at night (Song of Solomon 3:7). The two men who walked with Jesus after His resurrection were strengthened in their faith (Luke 24:13).

The Sermon
The Strength of the Believer

I. Be Strong in the Lord - Ephesians 6:10

The believer is admonished to be strong in the Lord. The believer should study the Word of God and pray. These two elements will provide faith in God, and make the believer strong in their service for the Lord Jesus Christ. The believer must have faith in the power of God's might.

II. Be Strong in Grace - II Timothy 2:1

The believer is encouraged to be strong in the grace of God. When the believer realizes the extent of God's love toward all man, and especially towards His children, then the believer becomes strong. Knowing that God loves the believer, and that the believer is on the side of God brings hope which makes the believer strong in the grace of God.

III. Be Strong Through Your Trials - Hebrews 11:34

The believer is challenged to be strong through his trials. Faith produces courage which makes the believer strong even during his trials. Trials can turn weaknesses into strength.

CHAPTER 57

THE NUMBER SEVENTY ~ "FRUITION"

The Source

The first mention of the number seventy is found in Genesis 5:12 where Caninan was the third descendent of Adam and had his first son, Mahaleel, at the age of seventy. The number seventy is associated with three common things: the age of men; the number of men in a group; and a period of time for a nation.

Concerning the age of men, Caninan was seventy when he had his first son (Genesis 5:12). Terah was the father of Abram, Nahor, and Haran, and he had Abram when he was seventy years of age (Genesis 11:26).

Concerning the number of men in a group, there were seventy souls in the loins of Jacob (Exodus 1:5). There were seventy elders of the nation of Israel (Exodus 24:1; Numbers 11:16). Ezekiel 8:11 mentions seventy men of the ancients of the house of Israel. Seventy disciples were sent by Jesus to preach (Luke 10:1,17).

Concerning the period of time for a nation, Isaiah mentions that the nation of Tyre will be forgotten for seventy years (Isaiah 23:15-17). Jeremiah 25:11 tells that the nation of Israel will be in captivity for seventy years (Daniel 9:2, 24).

The number seventy is also mentioned in Numbers 7:13 where one silver bowl of seventy shekels is spotlighted. Silver is a symbol of redemption. The number seventy is also mentioned by Jesus when He corrected Peter in the number of times to forgive someone. Where Peter thought seven times to forgive would be sufficient, Jesus told Peter that seven times seventy should be the goal (Matthew 18:22).

Seventy is the product of three prime numbers. Two times five times seven is seventy; where two is symbolic of witness; five is symbolic of grace, and seven is symbolic of perfection.

The Symbolism

The number seventy is associated with fruition. Fruition is the attainment of anything desired, the realization, accomplishment, the state of bearing fruit, or the period of time it takes for an event to be completed.

The gestation period for a woman carrying a baby to delivery is nine months. This is the time of fruition for the baby. The same is true in God's timetable in that He uses seventy for the number of men in a group or seventy for the number of years for an event to come to fruition.

The Sermon

Fruition

I. God Has a Timetable for Israel ~ Daniel 9:25 - 27

The future is being held in the hands of God. He knows how to place kings and remove them to fit His timetable for Israel. God has predicted in the Bible when some things will happen, and He will turn nations and civilizations to accomplish His purpose. His goal is to bring Israel back into His fold and have fellowship with them through Jesus, His Son, as He once did in the days of Moses. God has a timetable for the nation of Israel.

II. God Divided the Timetable in Three Parts ~ Daniel 9:25-27

God has divided the timetable for the nation of Israel in three parts. The total years will be seventy weeks, or seventy times seven years, or four hundred and ninety years. The four hundred and ninety year period begins with the commandment to rebuild the holy city. The four hundred ninety year period is divided into three modules: forty-nine years, four hundred thirty-four years, and seven years. The first section of forty-nine years was to be occupied with the actual completion of the streets and walls of the city in "troublous times" which is described in the books of Nehemiah and Malachi.

After the forty-nine year period, there was to be another period of four hundred thirty-four years before Messiah would appear and be cut off. The last seven week period will come after the rapture of the church and the start of the great tribulation. The revelation of the Antichrist will be given, and he will reign with a false peace until the middle of the great tribulation. At that time the Antichrist will proclaim himself as God and will desire the worship of the Jewish people. God's timetable for fruition of prophecy is seventy weeks and it is divided into three parts.

III. The Church is Blessed in the Age of Grace ~ Acts 2:21

The period between the sixty-ninth week and the last week of the seventy weeks is the age of grace towards the church. The people living in this time span will have a great opportunity for salvation, which is to simply call upon the Lord Jesus Christ to be saved. It may sound simple, but it is the blessing of living in the age of grace.

CHAPTER 58
THE NUMBER EIGHTY ~ "STRENGTH"
The Source

The number eighty is found twelve times in the Bible in the form of fourscore. A score is twenty therefore fourscore would be eighty. Some of the locations of the number eighty in the Scriptures are Exodus 7:7; Judges 3:30; II Samuel 19:32; I Chronicles 15:9; II Chronicles 26:17; Ezra 8:8, Psalm 90:10, Solomon's Song 6:8; Jeremiah 41:5 and Luke 16:7.

The Symbolism

The number eighty in the Scriptures is symbolic of strength. This is similar to the symbol of the number sixty, but there is a difference. Strength is the force or desire within a person. Strong is a temporary physical condition of a person or nation. Psalm 90:10 reveals that a person can live to be eighty if by reason of strength. If you have strength, you can live to be an older age. Moses was eighty years old when he returned to Egypt. It took inner strength to face the ruler of Egypt because Moses was a fugitive for killing an Egyptian. He worried that he was going to be arrested and executed if he went back to Egypt. It took inner strength to obey God.

In Judges 3:30, it took strength for Israel to subdue Moab. Moab had overrun Israel, defeated them, and had stolen from them. Finally, a leader came along, and through their army they subdued Moab and brought eighty years of peace. Peace was possible because they had strength to overcome the Moabites. Peace can be won through strength.

In II Chronicles 26:17, it took strength for the priests to resist King Uzziah. King Uzziah seemed to have things going well for him. He built towers in the desert and dug many wells. He had much cattle in the low country and in the plains. He had gardeners and vinedressers in the mountains because he loved husbandry. He was a farmer and rancher. He loved to see things grow so he developed the land. He had a host of fighting men that went to war by bands. Things were good for Uzziah. He had a large farming operation. He fed people, helped people, and raised up a good army of fighting men. Uzziah prepared his army by producing shields, spears, helmets, habergeons, bows, and slings. In Jerusalem he made engines invented by cunning men which were placed on the towers of the bulwarks to shoot arrows and stones. Uzziah had smart men invent the multi-arrow shooter. They could load multiple arrows, pull a crank back, and fire it. When they practiced, they knew exactly where the arrows would land. When the enemies got to that point, they simply fired the weapon and eliminated them. The bulwarks were the mounds around the walls of the city. The enemy could not come in with rolling engines and roll next to a wall to send warriors over it. The ground was not level around the city walls, so the enemy could not bring weapons against it. Uzziah seemed to have everything going for him. He became strong, and his reputation was such that no one would go against him. Unfortunately, when he was strong, his heart was lifted up to his destruction. This is victory, symbolized by the number seventeen, without wisdom, which is symbolized by the number nineteen. King Uzziah had the victory, but he did not apply wisdom to his life.

Uzziah transgressed against the Lord his God, and went to the temple of the Lord to burn incense upon the altar of incense. He was neither a Levite, nor a priest. He stepped out of his role as king, and stepped into the inappropriate role as a priest. He took incense, and was going to offer it to the Lord. Azariah the priest went in with fourscore priests of the Lord that were valiant men. These men had strength within themselves to go against a king who had authority to take off their heads. They, in essence, had an inner strength that helped them overcome their fear of the king and his mighty power. They had strength to do right in the face of wrong. Azariah and the fourscore priests went in the Temple after him, and they withstood Uzziah the king and told him that they were the only ones authorized by God to burn incense unto the Lord. They told Uzziah to leave the sanctuary, for he had trespassed. Uzziah was wroth, and he had a censor in his hand to burn incense. While he was mad with the priest, leprosy rose up in his forehead before the priest in the house of the Lord and beside the incense altar. Azariah and the priests looked upon him, and he was leprous in his forehead. They thrust him out of the Temple because the Lord had smitten him, and Uzziah the king was a leper until the day of his death. He dwelt in a separate house, being a leper, for he was cut off from the house of the Lord. Jotham, his son, was over the king's house, judging the people of the land. The number eighty is symbolic of strength, for this is the characteristic displayed by Azariah and the eighty priests who withstood one of the most powerful men in his time in the nation of Israel.

The Sermon

Strength

I. The Discovery of Our Strength ~ Nehemiah 8:10

In Nehemiah 8, the words of the Lord were read from a wooden pulpit, and when the people heard that they had done wrong, they started weeping. Nehemiah told them they should be sorry for their sins, but there was another element needed in their lives. He wanted them to be happy since they had realized their error and had turned from error. The joy of the Lord should be their strength. When the believer knows that God has taken care of his past, he cannot stay in a state of sorrow, but can rejoice in a God who forgives.

Believers should have joy knowing that their sins are forgiven, and that they have no sin. In heaven there is nothing condemning on their account. Nehemiah encouraged the children of Israel that the wall will be built and they are going to secure it. The temple is already built, and they are going to have a secure wall around the city. The joy of the Lord is their strength.

II. The Duty of Our Strength ~ Mark 12:30

In Mark 12:30, the duty of the strength of the believer is that the believer should love the Lord his God with all his heart, with all his soul, with all his mind, and with all his strength. This is the first commandment; and the second is like unto this, that thou shalt love your neighbor as yourself. There are four ways the believer is asked to love the Lord. The first is with the heart, which is the essence and being of who a person is. Then He said that the believer is to love the Lord with all his soul. The soul is the seat of all the emotions. The believer should love God with all his mind and should love him with his strength. This is the strength the believer has in their body. Two commandments are given, one toward God and one toward those around him, which forms a cross in our lives. There are four things God has asked of the believer, and four is symbolic of the world. Many a man has failed because he has a wishbone where his backbone ought to have been. God desires of the world that the people of the world love Him and obey Him.

III. The Disguise of His Strength

In II Corinthians 12:9, we have the Apostle Paul in a difficult situation. Paul was so close to God, that God revealed many things to him which no one else saw. The revelation of the catching away of the church was given to him. Because Paul had an abundance of revelations from God, he was given a thorn in the flesh which was the messenger of Satan to buffet him or strike him less Paul should be exalted above measure. Every time Paul tried to do something, the devil was always hitting him. A clear answer is never given to the church as to Paul's affliction. God did this so every believer could associate his personal problem with Paul's personal problem.

Paul sought the Lord three times that his affliction might depart from him. The Lord responded unto Paul by saying, "My grace is sufficient for thee, for my strength is made perfect in weakness." (II Corinthians 12:9). Paul stated that he would rather glory in his infirmities, that the power of Christ may rest upon him. This was the disguise of his strength. The believer's strength is made perfect in weakness. The disguise comes in the fact that when the believer is weak, then the believer starts depending more on God, and his faith rises. Strength in prayer is better than length in prayer. The strength of the believer is increased during adversity.

CHAPTER 59

THE NUMBER NINETY ~ "FRUITFUL"

The Source

The number ninety is only found three times in the Bible. The number ninety is not the same as the number one hundred ninety or three hundred ninety. These are totally distinct different numbers with their distinct symbolism. The Bible student should be aware of this fact and not try to associate the number ninety with the number one hundred ninety so as not to confuse the symbolism of different numbers.

In Genesis 5:9, Enos lived ninety years and begat Cainan. In Genesis 17:17 is the story of Sarah, who could not have children. The angel told them that she would be blessed with a son and she would be a mother of nations. Abraham fell upon his face and laughed. They named the boy Isaac, which means laughter. Abraham was one hundred years old and Sara was ninety. She was going to have a child, and it was to be a miracle. Abraham fell on his face, laughed and said in his heart, "Shall a child be born unto him that is one hundred years old and shall Sara who is ninety years old bear?"

The third time the number ninety is found is in the book of Ezekiel. The dimensions of the structure that would be built on the new earth are given. This building will be seventy cubits broad and the length will be ninety cubits (Ezekiel 41:12).

The Symbolism

The symbolism of the number ninety in the Scriptures is fruitful. Enos lived ninety years and was fruitful. He was ninety years old and had his first child, Cainan. The boy's name Cainan means nestling or a young bird in the nest. To have the first child at the age of ninety was not unusual in this era because men lived to be over nine hundred years of age. The first mention of the number ninety is associated with the idea of fruitfulness.

At the age of ninety, Sarah was told by God that even though she was past the child bearing age, she would have a son by Abraham. She became fruitful in her womb at the age of ninety.

In Genesis 5:9 there is mentioned that a man had a son. Then in Genesis 17 there was a woman who had a son. The idea of being fruitful, multiplying, doing the work of God, is seen in association with the number ninety. In most cases where the number ninety is found in the Scriptures, there is a picture of fruitfulness.

The number seventy means fruition. Fruition does not mean the same as fruit bearing or fruitfulness. Fruition is the length of time required to accomplish something, which is much different from being fruitful. One is a period of time to bear fruit, and the other is bearing the fruit. The number seventy deals with how long it took something to happen. God often uses the number seventy to bring about something that is completed. We know the period of gestation for a woman bearing a child is nine months. That is called the fruition, but the number ninety is bearing fruit. Fruition is nine months in this case, but the fruit is a child. Ninety is symbolic of being fruitful.

The Sermon

Bearing Fruit

I. Good Fruit - John 15:1-2

In John 15, Jesus said that He is the true vine and His Father is the husbandman, which means the gardener. He is the one who tends the garden. Jesus said, "Every branch in Me that beareth not fruit, He taketh away: and every branch that beareth fruit, He purgeth it that it may bring forth more fruit". The branch cannot bear fruit of itself. It must be attached to the vine. It is the grapevine that grows and bears the fruit. The new branches that come out in the spring are the ones that will produce new fruit. The gardener knows how to prune the vine.

Every Christian would bear good fruit if their life was less involved with the world. When somebody gets saved, there is a change inside. The Holy Spirit Who lives inside the believer, will start correcting, teaching, and showing the believer the correct actions. He will guide the believer in all manner of righteousness. If every believer will listen to the Holy Spirit, then he will make good decisions in his life, according to the admonition given to each church in the book of Revelation (Revelation 2:7).

II. More Fruit ~ John 15:2

When God sees there is fruit coming from a person's life, He will send a teacher, a preacher, an evangelist, or someone to teach the believer what to do and how it should be done. He will start molding the believer's life through the Word of God.

The believer bears more fruit when he obeys the commandments of God because it is his duty. There is a difference in bearing fruit and bearing more fruit. From the vine to the branches, it is difficult to tell where Christ ends and the believer starts. They look the same. This happened in Antioch because they were first called Christians in Antioch. People of Antioch, who had heard of Christ, observed the lives of the early believers and called them Christians.

III. Much Fruit ~ John 15:5

The believer bears much fruit when he obeys the Word of God because he loves God. Love is the strongest emotion. Paul said the love of Christ constrained him. He was stoned, three times beaten, and shipwrecked three times. Paul's love for God compelled Him to preach the Gospel. The believer can bear much fruit when he obeys the commandments of God, not because of duty, but because he loves Him.

CHAPTER 60

THE NUMBER ONE HUNDRED ~ "WHOLE"

The Source

The number one hundred is found around thirty-seven times in the Bible. The first mention of the number one hundred is found in Genesis 11:10. Shem was an hundred years old, and begat Arphaxad two years after the flood. He did not have children before the flood, and evidently he went on the ark at age ninety-nine, celebrated a birthday while on the ark, and two years later he had a son. The reason God placed Shem, Ham, and Japheth on the ark was to produce children so they could repopulate the earth.

Abraham was one hundred years old in Genesis 17:17 when his son, Isaac, was born. The Scriptures reveal the age of Abraham in this instance, to emphasize the fact that this birth was a miraculous one, which points to Christ. Of the thirty-seven plus times the number one hundred is mentioned in the Bible, the notable ones are these: In Leviticus 26:8, one hundred men filled with the power of God could chase ten thousand soldiers. In Judges 7:19, Gideon, at one time, had one hundred men. In I Kings 4:3, Solomon's daily provision of mutton was one hundred lambs per day. Every day they slaughtered one hundred lambs to feed his court and the royal guard. In I Kings 18:4, Obediah hid one hundred prophets from Jezebel. It is interesting that he did not hide ninety-five or ninty-nine, but he hid exactly one hundred. Isaiah 65:20 tells of a time that is coming when everyone will fill their days, and a child shall die at one hundred years.

In the New Testament, Jesus used the number one hundred to describe a man's sheepfold. He stated that a man who has a hundred sheep, will leave the ninety and nine and go hunt the lost. In describing the fruit that a Christian should produce, the highest goal was a hundredfold. The believers will bear fruit, more fruit, and much fruit. Jesus said that some believers will bear thirtyfold, some sixtyfold, and some a hundredfold.

In John 15:39, Nicodemus brought spices of about one hundred weight to anoint the dead body of Jesus. The Holy Spirit could have said that Nicodemus brought a large amount of spices, but He did not say that. He said that Nicodemus brought one hundred pounds of spices.

The Symbolism

The number one hundred is symbolic of wholeness or fullness. The word whole has the idea of a circle being completed and made full. The English phrase of giving it one hundred percent means to do it with a full effort or with your whole heart. When one says, "I love the Lord with my whole heart," it means there is no room in their life for other things.

Shem was one hundred when his first son was born. God put him on the ark to preserve the seed of man, and when he became one hundred years of age, it made a full circle because then he had a son. On board the ark were eight people, and that number is symbolic of new beginnings. From those eight people were all the genes and chromosomes for every person on the face of the earth. Using the gene pool from those eight people, God was able to produce all the nationalities that exist on the earth.

Jesus used the number one hundred to describe the sheepfold being full in Matthew 18:12. He was saying that when a man has a full sheepfold, he has one hundred sheep. When he counts them and there is only ninety-nine, he leaves them to find that one that was lost and brings it back so that the sheepfold may be full. That does not mean that a man would not have more than one hundred sheep, it is just that if a man had one hundred sheep in his sheepfold, then it would be considered whole or a complete flock.

The full goal of a believer is to have a hundredfold of fruit (Mark 4:8). That is the maximum or ultimate that should be in every believer's life. The believer should bring forth fruit, some thirtyfold, some sixtyfold, and some a hundredfold. The hundredfold is the ultimate goal that could be accomplished.

The Sermon

How To Love God

I. Love God with Your Whole Spirit ~ Colossians 3:23

In Mark 12:33, a man came to Jesus and asked which is the greatest commandment of all. Jesus told him to love the Lord thy God with all his heart, with all his soul, with all his mind, with all his strength. This is the first commandment. The second is like to this, he should love his neighbor as himself. There is no commandment greater than these. This man answered in a good way. He said if a man can love God with all his heart, soul, strength, and love his neighbor as himself, then that is more than all burnt offering and sacrifices. The spirit of the believer should not be directed toward the world but toward God. Jesus condensed the commandments of the Old Testament into two commandments.

II. Love God with Your Whole Soul - Psalms 119:145

The soul is the place of the emotions. This is one feature that makes humans different from animals. The animals have a spirit and a body, but they do not have a soul. Humans know the consequences of right and wrong. God has given every human a conscience. It makes humans aware that some acts are not right. Even man without God has a conscience that tells him whether things are right or wrong.

III. Love God with Your Whole Body - II Corinthians 3:1-2

The body or strength of the believer should be used as the epistle Paul mentions. The language of the believer should show the love of God in the heart and the hatred for the things of the world. When the speech, the body, and the strength of the believer are corresponding to the requirements of the Holy Scriptures, he will be recognized as a follower of Christ. The talk of the believer is important. The believer needs to talk for God and walk for Him.

CHAPTER 61
THE NUMBER ONE HUNDRED TWENTY ~ "ROYALTY"

The Source

The number one hundred twenty is found eleven times in the Scriptures. The first mention of one hundred twenty is found in Genesis 6:3, where God stated that man's days would be a hundred and twenty years. Notable are the mentions in Deuteronomy 31:2 of the death of Moses, and the one hundred and twenty in the upper room on the day of Pentecost.

The Symbolism

The number one hundred and twenty is symbolic of royalty. The location of this number in the Scriptures is an allusion to royalty. Genesis 6:3 contains a royal decree from God, the King of Kings, in declaring the limit to the years of man. In Numbers 7:86, there was one hundred and twenty shekels of gold in the royal spoons. In Deuteronomy 31:2 and 34:7 is a description of the funeral for Moses which was a royal tribute for a great man. In I Kings 10:10 - 11, Queen Sheba brought a royal gift to Solomon which was one hundred and twenty talents of gold, spices, and precious stones. The height of the porch of Solomon's Temple, a royal building, was one hundred and twenty cubits. Darius set over his kingdom one hundred and twenty princes (Daniel 6:1). No greater royal meeting was held than the one in the upper room (Acts 1:15) when one hundred and twenty disciples of Christ waited for the Promise. They were visited by a royal Personage, the Holy Spirit.

The Sermon

Royalty

I. We Have a Royal Redeemer - Matthew 2:1-2

The believer has a royal Redeemer for Jesus was born a king. The wise men from the East came to worship the King of the Jews. He did not then fulfill the office of His Kingship, but He is coming again as a King (Revelation 17:14; 19:16).

II. We Have a Royal Law - James 2:8

The believer has a royal law from God to love his neighbor as himself. This law could only be achieved if the believer loves God with all his heart, mind, soul, and strength. The royal law is the one that distinguishes the Christian believer from other religious groups in the world in that the believers of Christ have love one toward another.

III. We Are a Royal Priesthood ~ I Peter 2:9

The believer is a royal priesthood in his attitude. The attitude of the believer should be as a child of the King. The believer should be such that he would walk as a victor instead of a victim. The believer is a royal priesthood in his attitude of seeking his help from God.

CHAPTER 62
THE NUMBER ONE HUNDRED FIFTY ~ "EARTHLY AUTHORITY"

The Source

The number one hundred fifty is found six times in the Bible. The first mention of the number in the Scriptures is found in Genesis 7:24 where the waters prevailed upon the earth one hundred and fifty days during the great flood. The other times it is mentioned are in Genesis 8:3; I Kings 10:29; I Chronicles 8:40; Ezra 8:3 and Nehemiah 5:17.

The Symbolism

The number one hundred fifty is symbolic of earthly authority. In contrast, the number twelve is symbolic of divine authority. In Genesis 7:24 and in Genesis 8:3, the waters prevailed upon the earth. The water had the authority over everything on the earth except Noah's ark. In I Kings 10:29, the value of a horse was one hundred fifty shekels of silver. The horse is an earthly symbol of power. In I Chronicles 8:40, there are mentioned the one hundred fifty sons of Benjamin, where Benjamin means the son of the right hand, or authority. Ezra mentions one hundred fifty men who were the "chief of their fathers," indicating earthly authority over their clan. In Nehemiah 5:17, Nehemiah had one hundred fifty Jews and rulers eating at his table, which was the table of authority. The idea associated throughout the Scriptures for the number one hundred fifty is earthly authority.

The Sermon

Authority

I. The Authority on Earth of Faith - Matthew 8:13

The centurion was a man who knew authority. When Jesus said that He would go to his residence and heal the centurion's servant, the centurion recognized the authority that Jesus possessed. Understanding authority, the centurion told Jesus to speak the word only and his servant would be healed. He displayed great authority on earth by his faith. The believer of Christ has authority through faith in God over issues on earth.

II. The Authority on Earth of Prayer - Matthew 21:17-22

Jesus delivered a great lesson on prayer when the disciples commented that the fig tree had withered. Jesus told them that not only was the fig tree subject to the words of a disciple of His, but even a mountain could be moved and cast into the sea.

In his travels across Asia, Marco Polo recounts a story in his book, *The Adventures of Marco Polo*. Marco Polo was told this story by the Christians who lived in the Baghdad, Iraq, area in 1225 A.D. The new, reigning Muslim Caliph had schemed to trap the Christians using their Bible, and thus accuse them of being liars. The Caliph had read Matthew 17:20 and told his followers that he would trap the Christians with this verse. The Caliph assembled the Christians and told them they had ten days to move the mountain before them or embrace the law of the prophet Mohammed. After eight days of agonizing, praying, and fasting, a Bishop of exemplary life, dreamed of a one-eyed cobbler who could move the mountain. After locating the man in a neighboring village, the Christians persuaded the one-eyed cobbler to accompany them to pray for the mountain to be removed. The deadline arrived and the Christians carrying a processional cross before them, arrived at the foot of the mountain. The Caliph along with his guards also attended with the aim of destroying the Christians when they failed in their prayers. The reverent, one-eyed cobbler knelt before the cross and prayed to God in Heaven. At the end of his prayer, he cried with a loud voice, "In the name of the Father, Son, and Holy Ghost, I command thee, O mountain, to remove thyself!" As his voice quietened, the mountain moved and the earth trembled in a wonderful and alarming manner. Terror struck the Caliph and his men, but joy entered the hearts of the believers. God had answered the prayers of a common but devoted man. Authority on earth was demonstrated when this Christian prayed.

III. The Authority on Earth Hidden - Mark 11:27-33

When questioned on His authority to heal and teach the Word of God with power, Jesus asked the leaders of the Jews a question. He asked if the authority of John the Baptist was from heaven or of men. For fear of the people, the leaders stated they could not tell Him. Jesus responded with the same answer concerning His authority. His authority on earth was hidden from the religious crowd. The authority that the believer has in Christ is hidden from the world, but it is the most powerful authority on the earth.

CHAPTER 63
THE NUMBER ONE HUNDRED FIFTY-THREE ~ "EVANGELISM"

The Source

The number one hundred fifty-three is found one time in the Bible in John 21:11. It has four factors: one, nine, seventeen, and one hundred fifty-three. Four is symbolic of the world. Nine times seventeen equals one hundred fifty-three. Mathematically, one hundred fifty-three is a curious number for $1+2+3+4+5+6+7+8+9+10+11+12+13+14+15+16+17 = 153$. Also $(1 \times 1 \times 1) + (5 \times 5 \times 5) + (3 \times 3 \times 3) = 153$. The three digits of 1, 3, and 5 when added together equal 9 which is one of the factors of 153. Seventeen times nine equals one hundred fifty-three.

The Symbolism

The symbolism of the number one hundred fifty-three in the Scriptures is evangelism. The factors of one hundred fifty-three are nine and seventeen. Nine is symbolic of divine completeness and seventeen is symbolic of victory. When a person is saved, he has the victory along with divine completeness in his life.

In the beginning of His public ministry, Jesus asked certain men to follow Him and be fishers of men (Matthew 4:19).

According to John 21:2, there were seven men involved in the fishing expedition. Seven is symbolic of the number of perfection. At the end of his public ministry, Jesus tells the disciples to bring the fish which they had caught (John 21:10-11).

There were seventeen ethnic groups on the day of Pentecost when Peter preached (Acts 2:1). The symbolism of the number one hundred and fifty-three in the Scriptures is evangelism.

The Sermon
Evangelism

I. The Gospel Net - John 21:11

The disciples were invited to drop their fish nets and grab the Gospel net. Jesus had promised that they would be fishers of men. Preaching to the lost is like casting a net in a pool of water. A preacher will not catch every fish in the water, but some will be caught. The eternal security of the believer is illustrated by the fact that the net did not break. Preaching the gospel to the world is the last request from the lips of Jesus. The church is to go into all the world and teach all nations. The first step is to teach them God's plan of salvation.

II. The Cooperation in the Work - I Corinthians 3:6

Peter leads the other six men to bring the fish to shore. Paul stated that some would plant the Gospel and some would water, but God would give the increase. The cooperation of the believers and God is seen in the landing of the one hundred fifty-three fish on the shore. When believers work together along with God, then Gospel results will be achieved.

III. God Saves the Lost - II Peter 3:9

It was Jesus who told Peter and the other six men to drop their net on the correct side of the ship to be able to catch the one hundred fifty-three fish. It was Jesus who performed the miracle. Salvation is very simple. It is God who directs the believers to teach and preach the Gospel of Jesus Christ to the sinners. He is the One who can save the lost, not the church. The church may plant the seed of the Gospel and water the subsequent growth, but it is God who saves the lost, not man.

CHAPTER 64
THE NUMBER TWO HUNDRED ~ "UNACCEPTABLE"

The Source

The first mention of the number two hundred in the Scriptures is found in Joshua 7:21 where Achan stole two hundred shekels of silver. In II Samuel, Absalom cut his hair and it weighed two hundred shekels which is approximately six pounds four ounces. In Judges 17:4, Micah's mother purchased two hundred shekels of silver to form into an image to worship. The disciples had only enough money to buy two hundred pennyworth of bread (John 6:7). The little ship was two hundred cubits from the shore (John 21:8).

The Symbolism

The number two hundred symbolizes unacceptable in the Scriptures. Some Bible students symbolize the number two hundred as insufficient, which is the word used by the disciples in describing the amount of money they had to feed a large multitude. The problem with this symbolism is that it is not broad enough to adequately describe the idea associated with this number. Would it be best to describe the two hundred shekels which Achan took from Jericho as insufficient or unacceptable? Unacceptable is a finer, definitive word which describes the symbolism of this number.

Achan's sin of disobeying God's commandment to take nothing from Jericho was unacceptable (Joshua 7:21). Achan defied the tithe by taking for his own consumption the stolen items from Jericho. The first of ten belonged to God and this explains why God did not want the children of Israel to take a spoil of any of the items in Jericho. Jericho was the first city that the Israelis had attacked in a list of ten cities.

Absalom's heart was unacceptable to God, even though his beauty was exemplary (II Samuel 14:26). Micah's mother's actions were unacceptable to God (Judges 17:4). Two hundred pennyworth of bread was not only insufficient, but unacceptable, for the task ahead of feeding the multitude. A little ship two hundred cubits from the shore, was unacceptable in landing the number of fish which were caught (John 21:8).

The Sermon

Three Principles of Acceptability

I. Coveting and Stealing Are Never Acceptable in the Sight of God - Joshua 7:21

The things acceptable in the sinner's life are never acceptable in the sight of God. Achan saw items of value and took them, despite Joshua's warning to take nothing from Jericho. The sins of the flesh are unacceptable in the sight of God especially coveting and stealing. God took Joshua that the people should take nothing from Jericho, but Achan took items that were attracted to his eyes.

II. Outward Beauty is Unacceptable if the Inward Beauty is Nonexistent ~ II Samuel 14:26

Absalom had extraordinary outward beauty. The problem was that his inward beauty was nonexistent. The person with an inward beauty possesses a radiance not found in someone who is empty inside but may rather possess an outward attraction. God looks upon the heart and judges a person, not by their physical attributes or handicaps, but by the beauty in their heart.

III. Worshipping Someone or Something Else Beside God is Unacceptable ~ Judges 17:4

The mother of Micah led her family into idolatry. The two hundred shekels of silver which she used to make an idol for her home was unacceptable in the eyes of God. The making of idols in Micah's home led to worshipping false gods. The action of Micah's mother in making an idol led her to make clothing for her son, and made him a priest. Every action of Micah's mother was unacceptable in the sight of God.

CHAPTER 65
THE NUMBER THREE HUNDRED ~ "THE FATHER"

The Source

The number three hundred is found the first time in the Scriptures in Genesis 5:22 where it is noted that Enoch walked with God for three hundred years after the birth of Methuselah. In Genesis 6:15, God gave Noah the dimensions of the Ark which were three hundred cubits by fifty cubits by thirty cubits. The story of Gideon and his three hundred faithful warriors is stated in Judges, chapter seven. This passage describes how God reduced Gideon's army to three hundred men. From a group of misfits, David trained his army, and one man, Jashobeam slew three hundred of the enemy at one time (I Chronicles 11:11).

The number three hundred is found two times in the New Testament in the passages of Mark 14:5 and John 12:5. In Mark 14:5, while in the house of Simon the leper, Jesus was anointed with spikenard which was very precious. Someone in the group murmured against the woman who brought the ointment. The complaint was that the ointment could have been sold for three hundred pence and the money used to feed the poor. A very similar story is told in John 12:5 but the setting is in the home of Lazarus. Mary, the sister of Lazarus, anointed the feet of Jesus with a costly ointment.

The Symbolism

The symbolism in the Scriptures of the number three hundred is the Heavenly Father. The association of the number three hundred in Genesis 5:22 with Enoch's walk with the Heavenly Father is affirmed. Enoch walked with the Father for three hundred years after the birth of his son, Methuselah. For three hundred years, Enoch communed with God the Father and walked with Him.

The dimensions of the Ark built by Noah is given in Genesis 6:15 which was three hundred cubits by fifty cubits by thirty cubits. The number thirty is symbolic of the life (blood) of Jesus Christ, the Son of God. The number fifty is symbolic of God, the Holy Spirit, and the number three hundred is symbolic of the God the Father. In the dimensions of the Ark, which Noah built, are hidden the symbolism of the Trinity. Aboard the Ark was Noah who was not only father to Shem, Ham and Japheth but to the subsequent generations upon the earth. The father figure of Noah is undeniable. The number three hundred is symbolic of the Father.

Gideon is the main character in Judges, chapter seven. God informed Gideon that he had too many men in his army for the task ahead. Gideon's men may have claimed the glory for the success of the battle because of the abundance of ten thousand warriors. God instructed Gideon to give a simple test to downsize the army to the right size, which was three hundred men. God instructed Gideon to take the men to the water to drink. Only three hundred men bowed themselves upon their knees to drink water from their palms, while the other men prostrated themselves like dogs to drink. Gideon became a father figure to these three hundred men who kept watch as they drank water from the palms of their hands.

Two stories are given in the New Testament involving the number three hundred. A woman in Mark 14:5 enters the house of Simon the leper and anoints the head of Jesus with an ointment of spikenard. Complaints arose concerning the usage of the ointment. The argument was that the ointment could have been sold for three hundred pence, and been given to feed the poor. In John 12:5 a similar story unfolds. At the home of Lazarus, whom Jesus had raised from the dead just days before, Mary, the sister of Lazarus, anointed the feet of Jesus with a pound of ointment of spikenard which was very costly. A similar protest was made of Mary's action by Judas Iscariot. His accusation was that the ointment was not used in the most efficient way. It could have been sold for three hundred pence and given to feed the poor.

In these two verses, the body of Jesus is anointed on the head and on the feet. From head to foot, Jesus was and is God. The two women worshipped Jesus as God. He obeyed the Father in Heaven and submitted His will to the Father's will. Three hundred is symbolic of the Father in Heaven.

The Sermon

The Father and the Believer

I. The Walk with the Father - Genesis 5:22

Enoch walked with God for three hundred years. The believer's greatest opportunity is to have a close relationship with the Heavenly Father. The daily communion through prayer and the reading of His Word brings him close to the Father. When the believer walks with God, then the future expectation is to be with the Father in Heaven. Enoch experienced this phenomenon as he walked with God and was not, for God took him. The walk with the Father should be the desire of every believer of Jesus Christ.

II. The Work of the Father - Genesis 6:1

The Father, the Holy Spirit, and the Son work together to save the lost. How inadequate the Ark would have been if it had only two dimensions. With just the length of three hundred cubits and the width of fifty cubits, Noah would have built a raft. If he had built the Ark with only three hundred cubits long and thirty cubits high, then he would have built a wall. The work of the Father is in conjunction with the work of the Son and of the Holy Spirit. Just as the Ark of Noah required three dimensions to work effectively, the work of God involves the Father, the Son and the Holy Spirit. God the Father gave His Son to be the propitiation for the sins of the world. God the Son gave His blood for the payment of the sins of man. God the Holy Spirit performs the divine surgery along with the Word of God to give the sinner the new birth.

III. The Worship of the Father - Mark 14:5

The price of the worship of Jesus was judged not worthy to be three hundred pence by Judas. Jesus had lived a life which proved that He and the Father in Heaven were one (John 10:30). Jesus told the woman at the well in His discussion that worship must be in spirit and in truth, and that the Father seeks such to worship Him (John 4:23). One way to brag on grandparents or parents is to brag on their children or grandchildren. God the Father is worshipped whenever anyone places honor and respect upon God the Son.

CHAPTER 66

THE NUMBER THREE HUNDRED TWENTY-THREE ~ "DOMINION"

The Source

Only one time in the Bible is the number three hundred twenty-three revealed. In Ezra 2 the prophet Ezra described by census the number of people who traveled back to the promised land to occupy it. The number three hundred twenty-three appears in Ezra 2:17 where Bezai had three hundred twenty and three children. These were his children, grandchildren and possibly great-grandchildren who were living.

The number three hundred twenty and three is a product of seventeen and nineteen, a pair of twin prime numbers. A prime number is a number that has only two factors, itself and one. Twin prime numbers are rare in the number system, but mathematicians have proved there is an infinite number of prime numbers and twin prime numbers.

The Symbolism

The symbolism of the number three hundred twenty-three in the Bible is dominion. The name *Bezai* in the Hebrew means domineering. Three hundred twenty-three is the product of seventeen and nineteen. Seventeen is symbolic of victory and nineteen is symbolic of wisdom. The possession of victory and wisdom at the same time is rare. Noah had the victory over the flood, but did not use wisdom in drinking the fermented wine. He lost the victory (Genesis 9:21). David had victory over all his enemies and was the King of Israel. He did not use wisdom by inviting Bathsheba to his palace (II Samuel 11:2). Demas is commended on several occasions by the Apostle Paul and he had the victory. Demas evidently did not use wisdom because he forsook Paul for the things of the world (II Timothy 4:10).

The believer who has the victory (17) and maintains wisdom (19) will have dominion (323) over all worldly things. Job is a great example of a believer who had victory over sin and possessed wisdom throughout his trial. He was the domineering person throughout the book named after him.

The Sermon

Dominion

I. The Dominion Given to Man - Genesis 1:26; Psalm 8:5-9

Man has been given dominion over all upon the earth. God has placed man in control of all the animals and plants upon the earth. What God has created, then He has given man the dominion over those things.

II. The Dominion of God - Job 38:33-37

God has control over all the elements and laws which He has created in the universe. In communicating with Job, God through questions narrated that He has control over the physical world through His laws and actions.

III. The Dominion of Sin - Romans 6:14

The Apostle Paul gives the boundaries of the dominion of sin in the life of the believer. Because the believer is living under the grace of God, then sin has no dominion over him. If a person is living under the law, then sin does have control. The believer is entrapped by the sin and the law then condemns the believer. If the believer will live in the grace of God, then sin will have no dominion over him. The believer will have dominion over sin.

CHAPTER 67
THE NUMBER THREE HUNDRED SIXTY-FIVE ~ "CONSECRATION"

The Source

The only mention in the Bible of the number three hundred sixty-five is in Genesis 5:23 where the Bible says, "And all the days of Enoch were three hundred sixty and five years." The number three hundred sixty-five is clearly associated with the life of Enoch, where the phrase "all the days of Enoch" is given.

The Symbolism

The number three hundred sixty-five symbolizes consecration. Consecration is the dedication to some purpose. A person has one life and that life should be dedicated to one purpose, serving God. Enoch's life was dedicated to God. He had one life to live and he gave it to God. He was so consecrated to God that Genesis 5:24 says, "And Enoch walked with God: and he was not: for God took him." Enoch walked with God every day, not just the day that God took him.

Jude 1:14 states that Enoch was the seventh from Adam. Seven is symbolic of perfection. The seventh day in the pre-Christ calendar was a day of rest. Enoch walked with God 365 days each year till he was 365 years old. He was consecrated.

The Sermon

Consecration

I. Fellowship with God - Genesis 5:24

Enoch had fellowship with God by walking with Him on a daily basis. Walking with God is much different from talking with God or listening to God. Walking with God indicates the close fellowship Enoch had with God.

II. Faith in God - Hebrews 11:5

Hebrews 11:6 tells us that, "But without faith, it is impossible to please Him: for he that cometh to God must believe that He is, and that He is a rewarder of them that diligently seek him." Enoch did not simply awaken one morning and find himself walking with God, rather Enoch diligently sought God each day. Walking with God was not the producer of faith, rather faith produced Enoch's walk with God. The family of Enoch must have searched for him the day he left earth and went to Heaven, for Hebrews 11:5 says, "and was not found." Faith produces fellowship which results in a walk with God each day.

III. Fear of God - Jude 1:14-15

Consecration produces a reverent fear of God. The believer is not to be afraid of God, but to have a holy fear of God. The truly consecrated believer has a fear of God and judgment.

CHAPTER 68
THE NUMBER FOUR HUNDRED ~ "CHANGE"
The Source

The first mention of the number four hundred in the Scriptures is found in Genesis 15:13 which is a prophecy of the nation of Israel and is referenced in Acts 7:6. The children of Israel were treated in an evil manner in Egypt for four hundred years, and then God sent Moses. In Genesis 23:15-16, Abraham purchased a graveyard for four hundred shekels. In Genesis 32:6 and 33:1, Esau met Jacob with four hundred men. In I Samuel 22:1-2, four hundred discontented men gathered with David to form an army.

The Symbolism

The idea associated with the number four hundred in the Scriptures is change. In Genesis 15:13, there is a promise of change from the Israelis being bondservants to free men. There is roughly a four hundred year period between Malachi and Matthew's Gospel, and a change from the law to grace. In Genesis 23:15-16, Abraham purchased land to bury his wife, Sara, and there was a great change in the life of Abraham from that moment forward. In Genesis 32:6, Jacob is expecting his family to be killed by Esau with his four hundred man army, but there is a change from war to peace in their family. A group of unfit men become David's mighty men because of the change David made in them (I Samuel 22:1-2). Four hundred is symbolic of change.

The Sermon

Change

I. God Does Not Change, But He Can Make Changes - Malachi 3:6

Many things may change in society or in a nation, but God makes a promise to Israel and to the world that He does not change. Being omniscient, God knows the end from the beginning and does not have to change any position He has taken. He has the power, though, to make changes in a believer's life and in the direction of a nation.

II. Some Things Will Change In Our Lives - Philippians 3:21

God gives the promise to the Philippians that some things will change in the believer's life. The vile body that every believer possesses will be changed into a glorious body like the Savior's body. The greater changes, though, are the changes made in the believer's life as he conforms to the Word of God and through the work of the Holy Spirit in this life on earth.

III. God Takes His Time With Change Except the New Birth - II Corinthians 6:2

The believer will observe while reading the Bible, that God takes His time with changes in his life and in the affairs of nations. Israel remained in Egypt over four hundred years, but finally a change came. God operates on His own timetable and is not subject to any individual. God is willing, concerning sinners, to make the change in their life instantly.

CHAPTER 69

THE NUMBER FIVE HUNDRED ~ "EXCEPTIONAL"

The Source

The first mention of the number five hundred in the Scriptures is in Genesis 5:32 where the age of Noah is given. In Esther 9:6, the Jews killed five hundred of the soldiers who came against them in Shushan the palace. Included in the wealth of Job was five hundred yoke of oxen and five hundred she asses (Job 1:3) and the Scriptures proclaimed that Job was the greatest of all the men of the east. Ezekiel the prophet measured the Temple and found it was five hundred reeds long and five hundred broad (Ezekiel 42:20). In I Corinthians 15:6, Christ was seen of above five hundred brethren at once.

The Symbolism

The idea associated with the number five hundred in the Holy Scriptures is exceptional. Noah was exceptional in that from his life and the life of his wife would come their three sons. Combined with the lives of the three wives of their sons, there will be an exceptional lineage of all the persons upon the earth. In Esther 9:6, the Jews killed five hundred men who came against them in Shushan. These men were those who hated the Jews and sought to kill them. Instead, God gave an exceptional victory by destroying five hundred of the leading enemies of the Jewish people. Job had five hundred yoke of oxen and five hundred she asses (Job 1:3). These were exceptional trained animals that labored to plant the crops and aided in the harvest. Ezekiel measured the future Temple and found the dimensions of this exceptional building being five hundred reeds by five hundred reeds of measurement.

It is unusual to find a large number in the New Testament so this occurrence is exceptional. In I Corinthians 15:6 Christ was seen of above five hundred brethren at once. We would assume there were women present, but this number reflects only the number of men present for this occasion. This was truly an exceptional event. At this time, Christ was seen not just by a single person as in the case of Cephas, or of ten or eleven men in an upper room, but by five hundred men. These were exceptional men who were allowed to be present at this exceptional time. In contrast to the one hundred twenty in the upper room, these men will be the workmen of the Gospel. The one hundred twenty in the upper room were "royalty" or holy ambassadors of God. To spread the Gospel, God not only uses the royal ambassadors, but He uses these exceptional, five hundred men to proclaim the Word of God.

The Sermon

The Believer Is Exceptional

I. Exceptional in Potential ~ Genesis 5:32

The potential in the body of Noah was exceptional. From his genes and chromosomes would come the population of the entire world. God also used the makeup of his wife, sons and daughter-in-laws to accomplish this feat, but without the faith of Noah, there would not exist any descendants. Noah found grace in the eyes of the Lord. The believer is exceptional in the potential in his lives. The young convert possesses untold riches which God can use in the future.

II. Exceptional in Position ~ Job 1:3

The position of the five hundred yoke of oxen and the five hundred she asses may not appear to be exceptional but they were. Only two other animals were listed before the oxen and the she asses. These animals were essential for planting the crops, for watering them and for harvesting them. These animals were exceptional in their position in the great farming enterprise of Job. The believer is exceptional in their ability to plant the Word of God in the lives of others. Just as the she asses were used to water the crops and to harvest the crops, the believer is exceptional in his position of the promulgation of the Gospel.

III. Exceptional in Proclamation ~ I Corinthians 15:6

The five hundred brethren who saw the risen Savior, were immoveable in their stance of the Gospel. Someone may argue with one of these five hundred brethren, but their argument would fall short, because these five hundred brethren were eye witnesses of a risen Christ. The believer is exceptional in his proclamation of the Gospel also. Whereas the five hundred brethren were eye witnesses, the believer today proclaims the Gospel or good news of Christ by faith. The believer is exceptional in his proclamation of a risen Savior to a needy world.

CHAPTER 70
THE NUMBER SIX HUNDRED ~ "CHOSEN"

The Source

Noah was six hundred years of age when the flood was upon the earth (Genesis 7:6). When Pharaoh charged after the fleeing children of Israel, he took six hundred chosen chariots to capture them (Exodus 14:7). Six hundred men were selected for a special military mission (Judges 18:11 - 17). Samuel gathers together six hundred men to search for a new king (I Samuel 13:15). The weight of the head of the spear of Goliath was six hundred shekels of iron (I Samuel 17:7). David's chosen men were six hundred in number (I Samuel 23:13; 27:2; 30:9). King Solomon made targets of gold with six hundred shekels of gold in each target (I Kings 10:16). Solomon bought one chariot for the price of six hundred shekels of silver (I Kings 10:29). David purchased the threshingfloor of Ornan the Jebusite for six hundred shekels of gold (I Chronicles 21:25).

The Symbolism

The symbolism of the number six hundred in the Scriptures is chosen. Noah was chosen to propagate the human race (Genesis 7:6). Pharoah chose six hundred specific chariots to capture the escaping children of Israel. Samuel chose six hundred men to search for the new king (I Samuel 13:15).

David chose six hundred men as his initial military force (I Samuel 23:13). Solomon chose a chariot from Egypt and purchased it with six hundred shekels of silver (I Kings 10:29). The threshingfloor of Ornan the Jebusite was purchased by David for six hundred shekels of gold (I Chronicles 21:25). This spot was special. It was where Abraham offered his son, Isaac, for a sacrifice. It would become the site for the future House of Jehovah. It was a chosen place on the earth.

The Sermon

The Chosen Few

I. Mary Was A Chosen Vessel - Luke 1:30

Of all the women of the children of Israel, Mary, the mother of Jesus, was chosen to bear the Holy Child. This was a tremendous honor and was not given to any woman. Mary was a chosen vessel of honor of God. She was chosen.

II. Peter Was Chosen to Reach the Jews - Acts 2:14

Of all the twelve disciples, Peter was chosen by God to lead the evangelization of the Jewish people. He was chosen.

III. Paul Was Chosen to Reach the Gentiles - Acts 9:15

Freeing the disciple Ananias of his fears, God told him that He had chosen Saul of Tarsus to bear His name before the gentiles, kings, and the children of Israel. He was chosen.

CHAPTER 71

THE NUMBER SIX HUNDRED SIXTY-SIX ~ "CONTROLLER"

The Source

The number six hundred sixty-six is found four times in the Scriptures. The first instance is in I Kings 10:14 where the amount of gold that came to Solomon in one year was six hundred threescore and six talents of gold. In II Chronicles 9:13 the before mentioned weight of gold is alluded to again. In Ezra 2:13 the Scriptures say, "The children of Adonikam, six hundred sixty and six." The most familiar verse containing the number six hundred sixty-six is found in Revelation 13:18 where the number of the beast, the Anti-Christ, is stated to be the number of a man.

The Symbolism

The idea associated with the number six hundred sixty-six is controller. With the wealth that Solomon possessed, he controlled the kingdom and even the kingdom of surrounding countries. The wealth accumulated by Solomon came by the means of import tariffs on spices being transported from the east to the west. Solomon reportedly collected a ten percent tax on every product that travelled across the holy land. Traders, importers and vassal chiefs inflated the immense revenue of the country of Israel under Solomon's kingship.

In Ezra 2:13, the name of Adonikam is mentioned where Ezra gives a census of the people who had been captured by Nebuchadnezzar, and now their relatives were returning to the promised land. The children of Adonikam were numbered at six hundred sixty and six. The name Adonikam means "Lord of rising", "high", or "controller." This title would carry over to the Anti-Christ in Revelation 13:17 where he will cause all men, both small and great, rich and poor, free and bond, to receive a mark in their right hand or in their forehead to be able to purchase or sell any products. The Anti-Christ will be the ultimate controller of all men. Solomon controlled a major part of the world during his reign. Adonikam's name means *controller*. The idea associated with the number six hundred sixty-six is controller.

The Sermon

The Anti-Christ: The Ultimate Controller of Mankind

I He Will Control the Economy of the World - Revelation 13:16-17

The Anti-Christ will reach such an apex of power, that he will cause all persons, regardless of their societal standing, to receive a mark either in their right hand or in their forehead.

II He Will Control the Political Activity of the World - Revelation 13:15

By the means of his miraculous powers of giving life unto the image of the beast, the Anti-Christ through the false prophet will control the political operation of the world. He will dictate laws and enforce them. He will eliminate his enemies by death, and assume supreme power just as Hitler did in Germany before World War II.

III He Will Control the Souls of the World - Revelation 13:15

The Anti-Christ will control the souls of men and women of the world by requiring them to worship the statue of the beast. He will have the power to make the statue speak and assume life. By his tremendous, miraculous powers, the Anti-Christ will curse the souls of men and women who receive his mark and worship the living statue.

CHAPTER 72

The Divine Design of Digits

I. The Symbolism of the Prime Numbers

When God created the heaven and the earth in Genesis 1:1, He also created the laws of physics by which these created elements would operate. At the same instance, He created the number system and the mathematical implications of it. The animal kingdom and man is usually pictured with an even number. Man has two eyes, two ears, two arms and two legs. The plant kingdom is usually pictured with an odd number. Finding a four leaf clover is unusual. They are usually formed with three leaves. That is true of most of the plant kingdom.

The prime numbers are the building blocks of the number system. The number system goes from zero to positive infinity in one direction and it goes from zero to negative infinity in the other direction. Just as the prime numbers are the building blocks of the number system, the symbolism of these prime numbers explain the divine design of digits. The emphasis upon these symbolisms is by the design of God.

The symbolism of the prime numbers is:

Two - witness; Two is the only even prime number.

Three - complete;

Five - grace;

Seven - perfection;

Eleven - judgment;

Thirteen - rebellion;

Seventeen - victory

Nineteen - wisdom;

Twenty-three - death;

Twenty-nine - sanctification;

Thirty-one - cleansing;

Thirty-seven - mighty;

Forty-one - leadership.

The list above is only a partial list of the prime numbers. Mathematicians have proved that there are an infinite number of prime numbers. The symbolism of these prime numbers is very important.

II. The Symbolism of the Twin Prime Numbers

God spoke (Amos 3:3) through the Prophet Amos and stated, "Can two walk together, except they be agreed?" The twin prime numbers walk together in their symbolism. Twin prime numbers are prime numbers that are separated by only one digits with the numbers two and three being the exception. Two and three are the first twin prime numbers which respectively symbolize witness and complete. Three and five symbolize grace and perfection. Five and seven symbolize grace and perfection. Eleven and thirteen symbolize judgment and rebellion. Seventeen and nineteen symbolize victory and wisdom. Twenty-nine and thirty-one symbolize sanctification and cleansing. It is clear these twins walk together. Their meanings associate with one another.

The witness (2) of the Word of God will lead to completion (3) in a person's life. The completion (3) in a person's life (4) will lead to the grace of God (5). The grace of God (5) in a person's life will lead to perfection (7) one day in Heaven.

The judgment (11) of the lost will reveal their rebellion (13) against God at the Great White Throne Judgment. The continuation of victory (17) is secured by the wisdom (19) of the believer. The sanctification (29) and the cleansing (31) of the believer walk hand in hand. One is the cleaning of the inside of the believer and other is the cleaning of the believer on the outside.

CHAPTER 73

The Designer of All Things

Jehovah God is the divine Designer of all things. His handiwork is seen in the natural world, and as man has explored deeper, he has found the fingerprints of the divine Designer in the microscopic world. As man looks into the heavens, the craftwork of God is seen through the means of a telescope. Just as God designed all the eyes can behold, God has also designed the numbers to designate an idea. This has been revealed in the Holy Scriptures. The correlation of the prime numbers and those ideas associated with them is a revelation of a heavenly Creator. The connection of the twin prime numbers with each other also discloses the work of Jehovah God in an abstract discipline such as mathematics. The revelation of the symbolism of the numbers in the Holy Scriptures produces faith in the believer in God who does not randomly place words in the Bible. Jehovah God has a design of the numbers in the Holy Scriptures. The knowledge of the believer in the Word of God will increase when the symbolism of the numbers is investigated and faith will be increased.

PERSONAL DATA SHEET

John Randall Bell

Born on March 31, 1948, to Baptist Pastor John and his wife, Frieda E. Bird Bell;
Saved by the grace of God in August of 1961;
Education: Graduated with honors from Ringgold High School, Ringgold, GA, in 1966;
Graduated from West Georgia College, Carrollton, GA, with a B.A. Degree in mathematics, and a minor in physics in 1970;
Served two years in the United States Army in Europe;
Married to Mary E. Eidson on September 15, 1973,;
Graduated from West Georgia College, Carrollton, GA, with a M.A. Degree in mathematics in 1976;
Completed the Fifth Year level from West Georgia College, Carrollton, GA, with a degree in Administration and Supervision in 1980;
Called to preach the Gospel of Jesus Christ in July of 1980;
Called as Pastor of Riverside Baptist Church of Tallapoosa, GA, in September of 1980;
Called as Pastor of Harmony Baptist Church of Chattanooga, TN, in July of 1984;
Called as Pastor of Lupton Drive Baptist Church of Chattanooga, TN, in December of 1995;
Elected as Vice-President of Good Samaritan Baptist Missions of Villa Rica, GA, in October of 2012;
Doctor J. Randy Bell and his wife, Mary, have three children and seven grandchildren.